Writing Engineering Specifications,
2nd Edition

Writing Engineering Specifications, 2nd Edition

Paul Fitchett and Jeremy Haslam

Routledge
Taylor & Francis Group

LONDON AND NEW YORK

First published 1988 by E & FN Spon
Second edition published 2002 by Spon Press

Published 2023 by Routledge
4 Park Square, Milton Park, Abingdon, Oxon OX14 4RN
605 Third Avenue, New York, NY 10017

*Routledge is an imprint of the Taylor & Francis Group,
an informa business*

© 2002 Paul Fitchett and Jeremy Haslam

Typeset in Sabon by Prepress Projects Ltd, Perth, Scotland

British Library Cataloguing in Publication Data
A catalogue record for this book is available from the British Library

Library of Congress Cataloging in Publication Data
A catalog record for this book has been requested

ISBN 13: 978-0-415-26302-3 (hbk)
ISBN 13: 978-0-415-26303-0 (pbk)

Contents

Acknowledgements

Jeremy Haslam wrote this book in the late 1980s because he saw a need for it. There is still very little guidance to young (and older) engineers, and associated professionals, about how to write engineering specifications. When he asked me if I would write the second edition to bring the book up to date, I was surprised and flattered; but it was an offer that I could not refuse. The need to understand how to write specifications that fit in with contracting practices is as great now as it was when the first edition was written.

My special thanks go to Dr David Lax of Taylor Woodrow who reviewed the book with his usual vigour and goodwill and to Mrs Gaynor Haywood, librarian at Thames Water, who also very carefully reviewed the book. My thanks also go to Anne Dyde, who reviewed the chapter on grammar in the first edition and kindly did so again. My wife, Helen Fitchett, retired civil engineer and now a full-time secondary school teacher, was able to read the text from an informed position and from an 'outsider's' perspective. Where she found the time, in amongst the demands of the National Curriculum, I will never know.

The model of forms of contract in Figure 1.1 is reproduced with the kind permission of Professor Peter Thompson, who published it in 1980 in *The Organisation and Economics of Construction*. The model is widely used without any acknowledgement. It succinctly describes the principal effects of different forms of contract, and I am pleased to acknowledge the true originator. I should also like to thank David Glendinning for giving me permission to reproduce extracts from Thames Water Engineering's standard practice documents and the many other colleagues at Thames Water who have provided support. I should also like to thank David and Charles Publishing for the kind permission to use a quotation from *Great Ascents,* written by Eric Newby, which is the epigraph at the beginning of Chapter 2. The epigraphs in other chapters are taken from *Alice's Adventures in Wonderland* and *Through the Looking-Glass* by Lewis Carroll.

The Fog Index description is taken from *The ICE Essays. A Guide to Preparation and Writing*, published by the Institution of Civil Engineers. It was used in the first edition of this book and I am grateful to the Institution for renewing its permission.

Jeremy produced this book as a labour of love, and a beautiful job he made of it. I hope that I have not disfigured it too much with my updating and modifications.

Introduction

The term 'specifier' is used throughout this book as a device to describe anybody who regularly, or sporadically, finds that they have to write specifications for contract documents. There is not a huge amount of literature on the subject, and this book is intended as an easy reference and a simple guide for those who find that they are suddenly asked to make a written contribution to an engineering specification. Engineers are noted for their strong opinions and preferences, and I have tried to avoid giving opinions beyond those necessary for guidance on writing a specification.

The first two chapters in the book explain important factors that are the background to engineering specifications. It would be impossible to write a guide about how to write engineering specifications in isolation from this background. The following chapters look at the different stages in the mechanics of writing a specification, and finally the chapter on claims looks at what is effectively the key indicator of how well the specification has been written.

Significant changes have taken place and are taking place in the way in which engineering construction contracts are made, and these are to be welcomed. But the real change has been in the attitude of mind of many of the participants. Underneath all this, the basic principles of writing specifications are constant, and I hope that this is reflected in the book. Change in everything is necessary to keep minds lively and to stimulate interest. But I would encourage the specifier not to be carried away by any hype about what is the best approach (hype is unfortunately part and parcel of any activity carried out in our modern commercial system). The greatest asset a specifier can have is an open and enquiring mind. The greatest pitfall is to be drawn into the syndrome which Danny Kay and (later) Sinead O'Connor sang about – 'The Emperor's New Clothes'. Specifiers must always ask 'why?'.

The other major change that has taken place since the first edition was written is in the power of word-processing software. Gone are the days when a pair of scissors and a tube of paper glue were the specifier's best friends. Thank goodness for that.

1 Specifications in context

'What is the use of a book,' thought Alice, 'without pictures or conversations?'

Introduction

This first chapter looks at the work of the *specifier* (see the Introduction) as part of a commercial arrangement. It discusses contracts, contract and tender documents and pricing strategies. To know how to write a specification, the specifier needs to know how a contract is made and, should anything go wrong (in the legal, not the technical, sense), how the law of contract provides for remedies and assistance. The details of how contracts are actually prepared and how these remedies are obtained are usually left to lawyers.

No specifier is expected to know more than the basic information set out in this chapter, but it *is* very basic and it is not intended as a substitute for a legal textbook. Should the specifier wish to go further, there are a number of student texts on contract law as well as books on law for engineers, architects, and so on. However, the specifier should not be afraid of dipping into a legal textbook,[1] if only for the reason that most are beautifully written and many provide entertaining historical examples. The law of contract is very old and largely governed by events occurring in the horse-and-cart era – both appear frequently!

Contracts and bargains

A contract is a bargain, but not every bargain is a contract. Such a statement is a reliable guide to adopt when thinking of the background to the writing of any document that forms part of a bargain.

Over the centuries the development of the legal process – the general law – has recognized the need for people to make bargains and to seek the help of the state if the bargain is not honoured. However, in exchange for lending its support, the state, as the ultimate enforcer of the bargain, requires particular conditions (both procedural and substantive) to be met. Procedural conditions are those required by the state when it receives a request from a subject to enforce a bargain, and can loosely be thought of as methods of application to the courts or arbitration, and how those tribunals will hear the request. The substantive conditions are those contained in the common law (i.e. judge-made law) and statute. In these days the substantive law, as it also governs procedure, is always dominant, but this has not always been the case.

Properties of a contract

Not every bargain is a contract. In order to qualify as a contract, the bargain must have particular qualities:

1   an offer has been made;
2   that offer has been accepted;
3   there is an intention that the bargain be enforceable;
4   the parties making the bargain are qualified to do so under the law.

In the simplest terms, a bargain that exhibits the four qualities listed above is a contract and, as such, remains in existence until it has been performed or terminated.

The concept of offer and acceptance is based upon the legal notion of 'consideration' – the exchange of some benefit or burden for a corresponding benefit or burden, often in the form of a promise. In the normal course of events one party promises to pay, provided that the other party does something, although equally valid is the promise to pay provided the other party desists from doing something – a forbearance. The elements of a promise for a promise, or promise and forbearance in any combination, constitute valid consideration and make a contract binding in law. To promise unilaterally to do something for any one party (or even the whole world) for no apparent reciprocal action cannot be the basis of a contract; similarly, to forbear from doing something does not make a contract. Should a party wish to make a unilateral promise, he or she does so under a deed that needs no consideration. It used to be the case that a contract which was signed as a deed had to be 'signed, sealed and delivered' for it to be properly executed. Now, all that is required is that the parties make it clear, in the wording, that the agreement is intended to be a deed and that the correct signatures and witnesses are entered onto the contract. For a company this will normally require the signatures of two company directors or one director and the company secretary – legal advice should be sought. The other effect of signing a contract as a deed is to extend the period during which the parties may sue for their rights under the contract. The practical aspects of this effect are dealt with later in this chapter under 'warranties and guarantees'. Various general terms are used to distinguish contracts which are entered into as a deed from contracts entered into in the much more common way of offer and acceptance – 'special' and 'simple' are the usual respective terms.

In addition to the existence of consideration is the question of its value – and that question is ignored by the law. Consideration need have no value;

it merely has to exist. If a man wishes to let a house for the rent of a peppercorn a year, it makes a valid contract and consideration is said to have passed. If a building contractor enters into a contract to build something at a considerable loss, he or she cannot claim that no contract exists because of the lack of value (though the contractor may have other rights in contract). The last point is of central importance in engineering contracts in which there is no right to receive 'adequate' consideration (though there is a legal remedy against unjust enrichment), therefore any payment benefit has to be set out and received through the contract itself.

To many people there may appear to be some missing quality in all of the above: that a simple contract must be signed and be in writing. There is no legal requirement that a contract (except one relating to the sale of land) needs to be in writing. In fact, thousands of contracts are made every day on the Stock Exchange without any paper passing between the parties making them; the later 'contract note' is only evidence that a contract was made. And, as an extension to that idea, a contract can be varied by the parties making the original contract in the same way that the original contract was made. Indeed, many people have unwittingly made contracts by bargaining under the mistaken impression that, provided nothing is in writing, no contract exists. The point of written contracts will, hopefully, become clearer later in this book.

Returning to the four qualities of a contract, listed above, it is necessary to understand them all clearly to appreciate the process of turning a bargain into a contract. An offer must be made, and this is the easiest part. A person who offers to sell a car to another for £1,000 has made an offer. The recipient of the offer can do one of five things: he can reject it; accept it; make a counter-offer; make an enquiry; or ignore it. In order that the outcome is clear in a legal sense, the response to the offer must be specific. The necessary clarity is exhibited in these replies:

1 'Not likely/I can't afford that/No!' – these are clear rejections.
2 'OK/fine/here's the money/I'll give you a cheque/agreed' – these are clear acceptances.
3 'I'll give you £950/will you take £500 plus my hi-fi set?'– these are counter-offers.
4 'Can my dad see it first?/it's a bit pricey/will you take less?/what about part-exchange?' – these are enquiries.
5 'I'll let you know'/[no reply] – this is ignoring the offer.

Thus, (1) and (2) are both final – the first cancels the offer; the second gives rise to a binding contract. Once an offer has been rejected, it is no longer open for acceptance by the person who rejected it.

The counter-offer in (3) is merely an offer by the person to whom the original offer was made; it serves to reverse the offer–acceptance process. The counter-offer is therefore likely to produce any one of the five responses, including a counter-offer. This is the true bargaining process and is very common in any commercial transaction. The problem is knowing when the process has come to an end and a contract has been made.

The responses in (4) are in the nature of an enquiry into the attitude of the person making the offer. Unlike the counter-offer, an enquiry does not destroy the original offer, which is still open for acceptance when the enquiries are concluded. Thus, the making of enquiries is an uncertain process because the dividing line between an enquiry and a counter-offer is somewhat blurred.

To ignore an offer has no effect upon that offer, then, except that time (which may be a condition of the offer) is passing. The examples in (5) cannot be taken as acceptance or rejection. To state in an offer that the absence of response will be taken as acceptance is a futile gesture. The law does not recognize that a contract can be made by silence – except in rare cases in which the offer and response is already the subject of a binding contract.

An offer can be withdrawn at any time before it has been accepted, but after acceptance it cannot be withdrawn or modified. In order to be effective, the withdrawal of the offer must be communicated to the person to whom the offer was made. In some cases, a notice of withdrawal forms part of the offer, for example 'This offer is open for acceptance until noon on Monday, 22 October 2001', in which case no further notice is required. In addition, such an offer can still be withdrawn earlier, or the period of acceptance lengthened, simply by telling those to whom the offer was made. The term most commonly used for such a period is the 'validity period'.

Discharge of a contract

Once a bargain has been made, it must be performed by both parties in order that it can be considered as discharged. When both parties have carried out their obligations, neither can be required by the other to do anything else. Paradoxically, the process of cancellation, if allowed for in the contract, can be taken as part of its performance; the exercising of the right to cancel is merely performing the contract. To understand what constitutes performance, it is necessary to look at the terms of the actual contract to determine whether those terms have been performed.

Legal remedies

If the obligations under a contract have not been discharged either within the time allowed or in the manner required, then one of the parties is likely to have been at fault, that is in breach of contract. In such a case, the party who suffered has three remedies open to him or her under the general law:

1   damages;
2   performance;
3   repudiation.

The first remedy is the most common – obtaining money to compensate for the loss suffered. The amount of damages can be provided for in the contract as a pre-estimate of the damages that would be suffered in the event of a breach of contract. The term used to describe this type of damage is 'liquidated damages' (see p. 26). Alternatively, the damages can be left to be assessed (unliquidated damages), also referred to as 'damages' or 'damages at large'. Damages have to be related to the loss suffered to the extent that if no loss is suffered no damages are payable. The other point to bear in mind is that the loss is calculated as if the contract had been performed, not on the basis of it never having been made.

The second remedy of performance is rarely sought. Termed 'specific performance', it is a declaration by the courts or arbitrator that a party must perform its obligations or suffer a penalty imposed by the tribunal. As it requires a degree of supervision by the tribunal (for which it is not usually equipped), it is not often granted, and damages are awarded instead. The tribunal has thereby taken the line that money is sufficient compensation.

Repudiation, the third remedy, is the unilateral act of one party in cancelling the whole contract as a result of the other's breach. Although recognized in law as a valid course of action, the instances in which it can be successfully used are rare, coupled with the fact that repudiation itself may bring an action for breach of contract by the party who suffers the repudiation.

Contractual remedies

In addition to the remedies of damages, performance and repudiation under the general law of contract, the contract itself may contain some (or all) of the remedies open to an aggrieved party. Liquidated damages is one such remedy, and there are many others. In fact, the parties may agree to any remedies they choose, provided that they are not penalties. Parties to a contract are not allowed to seek punishment, only compensation for a loss. The common remedies are liquidated damages, offers of substitutes, reperformance of the unsatisfactory action to correct it and cancellation.

Limitation of rights

Whether or not a party is able to have remedies under the contract and under the general law of contract depends on the actual terms. Unlike consumer contracts, those in the business world may limit the rights of the parties to the terms of the contract only. Consumers are a special class of contracting person who have statutory rights which cannot be taken away by contract. Another class of person who may not be limited by contractual rights is one who suffers physical injury or death as a result of the actions of the other party to a contract in the course of its performance.

Warranties and remedies

The law provides remedies for those who have failed to receive what they were entitled to under the contract – common law rights. Additionally, there are remedies laid down in Acts of Parliament – statutory rights. But both rights in the general law and statutes are not, by their very nature, explicit. When buying a generator delivering 480 V, the buyer has no common law or statutory right to receive a generator with an error range of ± 10%, but there is a right to receive the item contracted for. If the contract specifies a generator with those properties, then the right is contractual.

The right to receive what has been contracted for is limited in time. The law (at present) states that a party has a right to sue within 6 years of when the fault was discovered or could reasonably have been discovered by the wronged party. This can be extended to 12 years if the contract is made as a deed. Although the law is somewhat fluid on the point, the trend is towards fixed liability periods.

Warranties and guarantees are contractual in nature and are intended to clarify the common law. How warranties (and guarantees – the words are synonymous) are intended to operate depends on which party is putting them forward. Typically, warranties offered by contractors are limited in nature, especially with respect to time. Any warranty that leaves the client with no redress for faulty equipment after a time which is less than the 6 years after the discovery of the fault has had the effect of limiting the client's common law rights. Any warranty which specifies the nature of the allowable claim (in technical terms) has the same effect. For example:

> X's industrial thermometers are guaranteed to ± 1 °C, valid for two years. No liability will be accepted by X for any loss or damage caused by errors outside the limit. X shall only be liable for a replacement thermometer up to the end of the warranty period, and shall have no liability thereafter. This guarantee substitutes all other rights of the purchaser.

Such a guarantee is commonly offered in industry. Note that it is absolutely limited in time to less than 6 years; the purchaser has no legal right to claim for damages if the thermometer fails to record the temperature accurately, even within the stated limits; the supplier is only liable for a replacement; and the purchaser cannot rely on his common law rights. Consider the effect when such a thermometer is bought by a business which stores germinating seedlings at a constant temperature, and the thermometer is inaccurate by ± 5 °C. However, if the thermometer was purchased by Mr Consumer for use in his greenhouse, then he would be protected by statute, which, although allowing for the time limitation, provides that the item must be fit for its purpose (it must actually record the temperature); the supplier is liable for the cost of the replacement thermometer including installation, delivery and return of the defective items with parts and labour; and any clause restricting his common law rights is invalid, thereby leaving the way open for consequential damages.

Obviously, the ordinary consumer is better off than the industrial client. This is intentional because it is recognized in law that consumers individually have limited bargaining power and therefore need statutory protection. Businesses, on the other hand, operate as equals in the bargaining process. The specifier must, then, be aware of what he is doing if he intends to obtain a specific warranty on the performance of any item. The advantage of certainty in a specific warranty, that is in the ease of proving a breach of warranty, has to be set against the limitations that such a warranty will contain.

Professional negligence

The ability to obtain or offer warranties is not restricted to contracts in which output is a *thing*. It is also available in those in which the output is a *service* (for example, the provision of security guards) or the product of a service (for example, a design). In cases in which a service is undertaken by a professional person (engineer, accountant, doctor, etc.), there exists the concept of 'professional negligence'. Those who give advice are here deemed to be expert because they hold themselves out to be so, and the client may expect a certain standard of advice that is normally available from that profession. But the subject is a complex one, and the law is far from clear. If the client wants to gain protection against inadequate advice, he or she will probably seek two things:

1   a clarification of what would be a breach of contract in the sense of failing to do work to a proper professional standard;
2   an assurance that the professional concerned could meet any financial liability concerned.

Therefore, a warranty may be drafted on the assumption that requirement (2) is covered by professional indemnity insurance because most professional people trade from a limited asset base.

The clarification of (1) is not, however, as easy as in the case of the thermometer but might well be along the following lines:

> The Contractor shall carry out the Design Services in accordance with the provisions of the Contract and good engineering practice and shall be liable for the consequences of his failure so to carry out the Design Services.

Alternatively, standard contract terms often describe the contractor's design responsibility as being that of 'skill, care and due diligence'. In any case, the legal position is complicated, and while the specifier should be aware of the basis of protection, the position is usually covered in the conditions of contract.

Intellectual property rights

The issue of intellectual property rights (IPR) is an important, yet complex area for contracting parties. IPR is a collective term that covers:

* patents;
* copyright;
* designs;
* trademark.

The area is extremely complex, and the law is developing continuously as scientific and medical boundaries are pushed back and as the worldwide web provides much greater access to information. The purpose of IPR is to provide incentives for invention and innovation – creative thought. The logic is that if the originator can profit from his or her invention then he or she will be more inclined to share that invention for the benefit of all of us. However, the logic of this argument becomes strained when intellectual property rights are granted on some discovery in human medicine, such as the 'human genome'.

In the engineering world, the specifier will come across copyright issues more often than not in two main areas:

1   copyright on the contract design;
2   copyright on third-party designs.

Copyright is an ownership right to reproduce drawings and documents, and it normally rests with the originator of the work. In an engineering contract, this would generally be the consultant or contractor. The client may well need to use the drawings and documents for any future purposes associated with that site and should, at the very least, obtain a free licence to use the copyright for these purposes. Third-party copyright can also be infringed in the pursuit of a contract. Generally, suppliers will be only too keen to provide drawings and information about their equipment if they are expecting to win the business. However, a spurned supplier might not be so happy if, say, CAD (computer-aided design) details of its equipment are used in the contract works. The client will need protection in the contract from such misuse.

New contracts for old

Since the parties to a bargain can make whatever arrangements they choose, it follows that they can substitute a new contract for a previous one – or, with the consent of the other party, they can replace themselves as the performer of the contract. This process is more common than would at first appear because companies change their identities and pass their contractual rights and liabilities on to others. For the process to be successful, legal advice is necessary.

Engineering contracts

Engineering contracts are not a special class of contract in any way. They must conform to all the legal rules applicable to all contracts, and all legal remedies are open to those who fail to receive the benefits to which they are entitled. However, engineering contracts do have some special features. First, they are invariably in writing, and second, they are often to a standard form of contract, nationally or internationally recognized. Third, they may contain provisions for a person, not a party to the contract, to exercise power under it on behalf of the client – i.e. the Engineer (see p. 27).

Why are engineering contracts in writing when this is not a requirement in law? The answer is that engineering contracts are complex bargains, the performance of which extends over a period of time. The committal to paper of the terms of the contract is the best way of ensuring that the terms are understood and remembered. In addition, because the terms are given in writing, the contracts are usually signed by the parties concluding the bargain.

Therefore, the purpose of contract documents is to provide evidence of the bargain, which is crucial to the enforcement of rights. Documents do not exist for the purpose of acquiring rights because these are part of the bargain and have their own existence. The distinction is of the utmost importance, and that is why it is crucial to draft all parts of the documents that form the contract with the utmost clarity. That, too, is why throughout this book there appears reference to the imperative need for:

1 consistency in terminology, particularly in the use of definitions and defined terms;
2 compatibility between all parts of the contract documents;
3 avoidance of repetition within and between parts of the contract documents;
4 the content of each part to be restricted to its function within the whole.

If, because of poor drafting, a party to the contract is able to prove that, despite the words of the contract documents, the bargain was really something else, then there must have been a failure in the presentation of the bargain in writing. An example of such a misunderstanding might be found in the construction of a tunnel:

> The contractor thought that he was to construct a tunnel in soft rock, and the client thought that he was paying for a tunnel in any sort of rock: the documents were unclear on the type of rock.

Clearly, then, the written version of this bargain is incorrect. Such an example may seem crass, but all specifiers can probably think of similar ones. Contract documents should therefore evidence the bargain and show at least what one party is to produce (i.e. construct, manufacture, write or draw) and the other party is to pay.

In addition, the documents usually set out the consequences of the failure to produce or pay, and sometimes to change what is to be produced or paid. The documents also allocate between the parties the risks inherent in the production. Risk allocation is a very wide topic but includes financial risk (e.g. currency fluctuation), damage to other people's property, interference with patent rights, reliability of data, physical risks to personnel (e.g. the Health and Safety at Work Act 1974), physical risk to the property of the parties, consequential loss (e.g. loss of profit due to the failure of the produced item), and so on. A document that does not deal with risk allocation, or omits a particular risk, means that the parties are relying on the common law in the event of a dispute over who is to bear the risk-related loss.

Contract documents

In engineering contracts the documents follow a common (though not a fixed) pattern. The pattern is best described by listing the contents of each part under its common name:

- *Conditions of Contract* or *Terms* or *Articles* of *Agreement:* sets out the undertaking to pay and the undertaking to produce; makes clear the offer and acceptance; allocates the risks; lays out the time to produce; and sets out the consequences of failure to pay or produce and the rights of the parties on other matters the parties consider relevant.
- *General Specification* or *Scope of Work:* sets out the scope of the work to be performed and the administrative procedures to control the performance of the work.
- *Schedule of Rates (Prices)* or *Bills of Quantities* or *Bill of Rates:* sets out the method of payment, the amounts payable, the currency of payment and any formulae relating to the payment amounts (escalation, currency fluctuation).
- *Particular Specification* or *Technical Specification:* sets out the technical details of what is required, and usually includes the drawings.
- *The Contract Agreement:* here it is often useful to preface all the documents with an agreement which sets out the documents forming the contract and containing the precise identities of the parties and their signatures; the only time that a formal agreement complete with seals is required is when it is necessary to turn a 'simple' contract into a 'deed' in order to extend particular legal rights; deeds are usually only arranged with the help of lawyers.

There may also be (though the practice is to be deplored) appendices consisting of exchanges of letters, e-mails, minutes of meetings and even parts of the tender.

Of course, the documents that appear in the above list depend for their exact titles on the particular house style of the organization that is preparing them. In the end, it does not matter what the constituent parts are called, or even if the contract documents are divided into parts at all. It is more important to include the details of the bargain than to conform to a rigid policy of referencing of parts.

Tendering and negotiation

What follows under the next five headings is a description of the tendering process for engineering contracts. The reader will realize from points made elsewhere in this book that there are alternative processes by which clients might form contracts with contractors – variations on a theme. The method that follows can be regarded as the basic default process and the principles that are contained in it can be adopted and adapted by clients into any variant process.

The final contract is the end of the offer–acceptance chain. In the beginning there must be an offer, and in the end there must be acceptance. The final offer and acceptance may be somewhat different to that originally envisaged because of the commercial realities of life. Basically there are three ways of making an offer:

1 by issuing a simple offer which allows a single identified party to accept or reject;
2 by offering to negotiate a bargain with a particular party;
3 by issuing an invitation to tender to one or more parties (the latter not necessarily being previously identified) for them to submit an offer to carry out the work.

Example (1) is straightforward and its requirements are just the simplest elements of (2) and (3). Negotiation in example (2) is a complex process of offer and counter-offer, again a version of (3) but aimed at a single party only. The issue of the tender in example (3) contains all the elements of the process from offer to acceptance. However, it is slightly complicated by the fact that the issue of the tender is not the making of an offer but the seeking of an offer from the tenderer. The client will then accept (or reject) the offer made by the tenderer. It is, then, like the man who asks another: 'Will you make an offer for my car?' The resulting offer, if it comes, can be accepted or rejected. In a similar way, the person who goes into a supermarket and chooses an item is offering to purchase the item (usually for the amount shown on the label), the acceptance being concluded at the till. If, at the till, a bargaining process takes place (for example, because of a defect in the goods), then this is merely an example of the realities of any commercial deal. The resolution becomes the offer and acceptance.

Tender documents

In order for anyone to react to an offer, he or she must have information. Information must be sufficient for the bargaining to be concluded successfully. In certain circumstances it is illegal to create a contractual relationship based on unfairness. Therefore, it is best to be quite open in the information given, but allow the other person to analyse it and draw their own conclusions. Thus, the issue of a tender should contain, at the very least:

• basic information on the work to be performed;
• how the work is to be paid for;
• how the tenderer is to present the tender;
• how the rights of the parties are to be governed.

The tender is often accompanied by various questions, both commercial (for details of the tendering company) and technical (what technical criteria the tenderer is offering). The number and detail of the questions depends on the extent, if any, of pre-qualification of the tenderers; pre-qualification is the process of investigating suitable companies prior to the issue of the tender. The tender documents usually consist of the following parts:

• instructions to tenderers;
• conditions of contract;
• scope of work;
• payment terms.

These all (except the instructions) have a direct parallel to the documents that make up the contract documents. The different terminology used to describe the various documents is not intended to be an example of sloppy drafting, but rather an example of the various descriptive terms applied.

The instructions to tenderers will not form part of the contract documents. The instructions merely serve to explain to the tenderers what they have to do in order to submit a valid tender. Matters commonly addressed are the following:

1 a brief description of the work and the payment terms;
2 the length of the tendering period;
3 who is the contact point in the client's organization;
4 how many copies of the tender to submit;
5 where to send the tenders;
6 how queries arising during the tendering period will be dealt with;
7 arrangements for visiting the site (if it is a construction tender);
8 what will be the evaluation criteria, time for award and award procedures (in some cases).

There is no set list, and the client can put as much explanation in as he or she cares.

Form of tender

The 'form of tender' is exactly that – it is a form with blanks which the tenderer generally copies on to his or her own letterhead, fills in the blanks and then signs. The form is used as a means of presenting a complete offer by referencing the documents sent to the tenderer and submitted as the tender. It will say, at the very least, that:

1 the tenderer has read and accepts the contents of the tender documents;
2 the tenderer offers to carry out the work in accordance with the sum, prices, or rates that he or she has entered in the tender;
3 the tender will remain open for acceptance for a specified period.

Examples of forms of tender can be found in almost all the standard forms of contract. Notwithstanding its formal appearance, the form of tender has no other properties than those contained in any other offer and acceptance routine.

Tendering process: from issue to receipt

The period of tendering lasts from the time that the documents are issued to the tenderers, or made available to them, or an advertisement is published, to the time that a contract is signed with the successful tenderer or the client cancels the whole process. This heading covers the period from when the tenders are made available to tenderers to when they are returned to the client. The most important point to make is that the length of the tender period should be appropriate to the complexity of the information that is being requested from the tenderers. Consequently, tenders that involve the submission of design proposals will require a longer tender period than those that are for construction services only. In addition, tenders that are based on a lump sum price will require a longer tender period because the tenderers have to be as certain as possible about their price to be able to submit their best price. If they are uncertain, then they are at risk (see Figure 1.1, p. 22) and the wise tenderer will add in a contingency to the price. The unwise tenderer might not, and this may well lead to an unsatisfactory contract that is plagued by claims. The typical range of tender periods is 3–4 weeks for a medium-sized construct-only contract, priced on a bill of quantities, to perhaps 12–14 weeks for a design and construct contract, priced on a lump sum basis.

Clients must take account of the fact that tendering is a major cost to contractors and structure the tendering process accordingly. The tender period and the number of tenderers must be appropriate to the client's needs and to the size and complexity of the project. When the economy is buoyant and contractors are in demand, they will naturally decline to tender for those clients who involve them in unnecessary costs in the tendering process. The client decides who is to be allowed to tender. Invitations to tender can be classed as follows:

1   Open tendering: used by many government agencies but few commercial organizations – literally anyone who wants to tender can do so.
2   Closed or selective tendering: used by many government agencies and most commercial organizations – the client chooses tenderers on some criteria of his or her own.
3   Deposit or bonded tendering: used by many government agencies – anyone who is willing to deposit a sum of money or a tender bond with the client will be allowed to tender. A tenderer refusing to sign a contract based on his or her tender forfeits the deposit or bond.

Another factor to be considered is whether the names of those who are tendering are to be made public. Public and private tendering is used by every organization and both philosophies have their adherents. Governments who have a consideration for public accountability use public tendering, whereas commercial organizations worry about confidentiality. However, in practice, 'secret' tender lists are not secret for long because any contractor tendering for a major contract will need the help of subcontractors, and so a sort of subtendering process will be run by the main contractors. In that case, there may well be advantages for the client in publishing a list of tenderers, so that a reasonable selection of subcontractors can be attracted. Many clients now make such information available publicly on their websites.

Once the tendering process is under way it follows the procedures set out in the instructions, although in all likelihood there will be little contact between the client and the tenderers. If there is, this consists of queries regarding the tender documents or visits to the worksite. On the basis that the client will want to treat all tenderers in an equal fashion, queries regarding the tender documents will be answered by sending a circular advice note to all tenderers, quoting the questions asked and the client's responses.

Visits by tenderers to the worksite are much more difficult to handle than written questions. Depending on whether the number of tenderers is large or they are all local, or there is a public or secret tender list, the client will have to choose the method of managing the visit accordingly. Perhaps the best way is to allow all tenderers an equal chance of asking questions during the visit or afterwards, but only to answer technical questions on the spot. Contractual or commercial questions can be submitted for written replies. Provided that the client deals fairly with tenderers, such visits are usually successful. Clients believe that tenderers in visits with other tenderers learn far more than they would if they visited the site separately – contractors have a different view. However, a single site visit for all the tenderers has the advantages of demonstrating the client's impartiality and of minimizing the client's tendering costs.

Tendering process: from receipt to award

A contract is a bargain, and that part of the tendering process that runs from the receipt of tenders at the client's office to the eventual award is probably the most important. In this phase the bargain is concluded. If it is carried out efficiently, the work will start on time. If it is carried out accurately, the parties will have a sound bargain, and recourse to the contract documents will resolve any problems that arise.

The question that has to be answered after the tenders have been received is: have the tenderers made offers that are capable of acceptance? A question so wide must often be reduced to a series of specific queries that have to be answered by specialists. For any one tender the question can be split into three, as follows:

1 Is the tender procedurally correct (i.e. delivered on time, correct number of copies, properly signed)?
2 Is the tender commercially attractive (without adjustments for missing items)?
3 Does the tender meet the requirements of the specification?

The first question is quickly dealt with; the other two form the main thrust of the evaluation process. It is generally advisable for the commercial evaluators to have sight of the whole tender, whereas technical evaluators should not see the commercial sections until they have produced at least a 'first pass' at their efforts. The reason for this is that apparent commercial ranking can easily influence a technical assessment. In many areas, the tendering process can be examined at a later date to check that the procedure followed has complied with company procedures (it may be that the tender assessment process is an integral part of a company's quality system) or perhaps with some external regulations such as the EU requirements for procurement by public utilities. In any event, the agreement of an assessment matrix with predetermined weightings can be a useful aid to this first stage of assessment. The commercial evaluators should see the technical parts, not because they can evaluate them, but because it is common to find commercial qualifications littered among apparently technical submissions.

The first evaluation aims to reveal errors and inconsistencies in the tenders, and will probably lead to a list of queries. If the number of tenderers is large, those tenders which are the most unattractive will probably be shelved. The decision to proceed with a shortlist is a management decision of great importance. However, it is not practical to carry all tenders through the evaluation process, and organizations set criteria for discarding tenders as the evaluation progresses.

Assuming that the legal and commercial analyses follow a parallel course, the evaluators of the specification will need answers from the tenderers. This is the purpose of a process sometimes called 'clarification' – as opposed to negotiation, which may come later.

Clarification of tenders, either in writing or at meetings with answers confirmed in writing, is purely designed to see whether the tenderer is capable of meeting the specification. It is not a process that is intended to enhance the specification within the price tendered. If the clarification reveals anomalies in the client's documents, and those anomalies are exposed in varying degrees by the tenderers, then the tendering process is in danger of becoming unwound. The essential factor is to maintain comparability of tenders, and this may mean that partial retendering is necessary. However, retendering is generally to be avoided because of the damaging effect on the programme. Less damaging is a partial retendering in which some or all of the tenderers are issued with a new version of that part of the tender documents which needed modification. The tenderers are then requested to confirm or amend their tenders accordingly. When the clarification (or retendering) process is complete, the tenders should be comparable, and then it is possible to choose the most attractive and proceed to the award of contract.

Whether the final stages in the process are carried out with more than one tenderer is a management decision, based on a perception of the market forces that influence the tenderer's desire to sign a contract. Since the tendering activity takes time, the tenderer's position may have changed, for example he or she may now have other tenders about to result in awards. Since the final stage is the production of a document that sets out what both parties intend to do, the need for accuracy in recording those intentions is paramount. Unfortunately, this stage commences with a decision in principle to proceed with a particular tenderer. Thus, any delay is viewed uncompromisingly by those who 'want to get on with it'.

When carrying out the compilation of the contract documents, it is not advisable to accept the tender unless it has literally no qualifying statements. If there has been a clarification process, there may now be negotiation to resolve how many of the clarifications can be accepted and how they will be incorporated in the documents. The specification will have to be amended – and not by adding letters, e-mails and minutes of meetings in some disjointed appendix. Such papers are never drafted specifically for inclusion in any document, so the points they contain are usually unclear (even if apparently clear in people's memories). This practice used to be common when it was difficult to make amendments to lengthy documents that had been typed or put together in some 'user-unfriendly' word-processing software. Now there is no great excuse not to put the agreed amendments into the body of the original tender document in order to create the contract document. Pains must be taken to identify all points of qualification in the tenders; these are resolved by rejection, amendment or acceptance. The drafter of the final documents incorporates all the acceptable points into the specification.

The object of the process is to produce a self-contained document and one that forms the entire bargain. This is often achieved by the insertion into the more legal parts of the document of a phrase such as: 'All previous communications between the parties on the matters regarding this contract document are superseded by the provisions of this contract document.' This is intended to make the bargain between the parties certain, and rules out any future reference to the tender or subsequent exchanges prior to the award over the precise intent of the contract. This is an important legal point, and most certainly applies to the specification. A further aspect to this point is the use of the word 'tenderer' in the enquiry document. The specifier should avoid its use completely in the documents that will form the contract, and the only place it should appear is in the instructions to tenderers. The tenderer will not be a defined party in the contract.

Aside from the actual specification, one of the most complicated series of points to incorporate is the tenderer's statement of his or her method of working (if asked to supply one). As part of the tender instructions it is easy to include a requirement that the tenderer shall supply the method of working, but not so easy to know what to do with it. It is necessary to ask – was the intention purely to ascertain whether the tenderer is capable of doing the work or was it to commit the tenderer to a particular method? If it was the former, then it should not be included in the contract; if it was the latter, then the method must be redrafted, so that it can be included.

At the conclusion of the tendering process is the award of the contract to the successful tenderer. The award is simply the formal offer by the client to the tenderer, which the tenderer then accepts by signing. The signatures of both parties indicate that the offer–acceptance routine is complete. The contract is made.

Hierarchy of documents

As an important aside on the nature of the contract documents, engineering contracts are usually made up of a number of separate documents which need to be tied together in such a way that they do not contradict one another (the purpose of this book), and if they do, then there is a clear order of precedence by which interpretation will be carried out. As a general rule, documents vary in importance according to the chronological order in which they were written or came into existence (later ones being the more important) or by the relative importance of their content or by some agreed order of precedence. In engineering contracts it is usual to state what that order is, for example:

1   the contract agreement and conditions;
2   the pricing section;
3   the specification;
4   the drawings.

Lists vary widely, but it is usual for the specification to be subordinate to the contract conditions, though the relationship with the drawings and the pricing section is much more difficult to be subject to any generalization. In fact, the hierarchy can be upset by the wording of the documents themselves, by the use of phrases such as 'notwithstanding previous references..., or indeed by a particular national law (for instance the Tender Regulations of Saudi Arabia).

Letter of intent

Once a decision has been made as to whom the contract is to be awarded, pressure from those who want to 'get on with it' may well become irresistible. The letter (or fax or e-mail) of intent is a method of identifying to all tenderers, to the chosen tenderer, and even to the general public, that the client has virtually concluded the evaluation of the tenders and, in principle, has made a decision. However, it is clear the client has not agreed with the tenderer the exact form that the bargain will take, otherwise contract documents would be available for signature.

The precise stage at which the letter of intent is appropriate depends wholly on the client's commercial policy. The only sure fact is that it is a vastly inferior substitute for the agreement to the contract. However, it may have a value aside from the substitution process, especially as a political expedient or in connection with tax minimization whereby a written agreement is required by a certain date. It may also serve to reserve facilities in the tenderer's works (shipyards, steelmills, etc.) or allow more time to negotiate subcontracts.

A letter of intent exists in only one form – as notification of one party's intention to do something. Because of its unilateral nature, it is not part of the process that leads to a binding contract. Should it require the recipient to accept its terms and signify that he or she will enter into a binding contract if required, then it merely extends itself to become an agreement to agree, which is no contract. As a result, a letter of intent has no legal force and cannot be taken as binding on either party, and it has been known to have been revoked by the issuer or ignored by the recipient without the possibility of any legal redress.

However, a so-called 'letter of intent' does exist in a modified form, as an interim arrangement. An interim arrangement is useful if the need for a letter of intent is to enable the successful tenderer to start work in the knowledge that he or she will eventually be awarded the contract, or at least be able to work in the interim and be paid.

In order for an interim arrangement to be effective as a binding arrangement, there must be certainty about its terms. That is, it must have the *properties* of a contract. Therefore, its drafting needs care, so that there is reference to those parts of the tender upon which agreement exists. The omission of contentious items, then, is essential to maintain its contractual nature. Nevertheless, the omission of contentious items underlines the undesirability of the whole process, because it is those items forming the basis of *disagreement* which may prevent a contract coming into effect to supersede the interim arrangement. As a result, one party may be financially, or otherwise, exposed while the interim arrangement operates. Of course, it should be obvious that the issue of a letter of intent will have a detrimental effect on the perceived bargaining strength of the client and the eventual ability to achieve the terms desired.

Drawings

Drawings are usually part of the tender documents and therefore will be relied on by the tenderers in the completion of the tenders. Drawings have no definition requirements that differ from those applicable to text; drawings define in a different form. Drawings must go through the same stages as other parts of the tender documents. Tender drawings will be commented on by the tenderers in different ways, and points on drawings will have to be clarified and alterations made if necessary, and thereafter negotiated, priced and agreed. The final contract documents must contain the pictorial evidence of the bargain, so will have to be amended prior to award. All too often, drawings are ignored by the client in the process of tendering, but they are rarely ignored by the tenderers.

Throughout the performance of the contract it is often the case that drawings are developed into shop drawings or construction drawings and thereafter into as-built drawings. Definitions of such drawings will need to be considered in the contract documents. However, as applies to the other terms of the contract, neither party can unilaterally revise the drawings to suit itself – that would be changing the bargain. Drawing revisions can only be carried out according to the terms of the contract.

Degree of definition

The choice between methods of payment is primarily dictated by the degree of definition that the specification has been able to achieve. From the detailed, complete specification to the conceptual, the choice runs from fixed-price lump sum to fully reimbursable. In addition, the following must be considered:

1 the degree of responsibility to be placed on the contractor by the client;
2 the level of control the client intends to exert on the contractor;
3 the risks and technology involved;
4 fair reward and motivation.

As usual, the middle ground is the hardest place in which to choose a suitable payment form.

Another factor which intrudes is the so-called 'state of the market', a consideration born of the idea that in lean times contractors will be prepared to fix prices for a less well-defined workscope than in times when work is plentiful. This is a dangerous delusion, and the specifier should always beware of others trying to raise the notional level of definition to a higher plane than that warranted by the facts. Specifiers must accept responsibility for recognizing the level of definition of their output. The correct way to approach the market is to obtain the most suitable form of payment for the level of definition; it is a mistake to let the market dictate the form of payment. Contractors may be willing and able to take risk, based on the specification

– but they are not able to price what may or may not be in the client's mind. Bailing out a contractor or receiving huge claims reflects no credit on anyone.

Forms of payment

The form of payment to the contractor is an integral part of the overall contract strategy. Although payment often reflects the risks and responsibilities the contractor is expected to take, it should be remembered that discussions on payment are not sufficient to formulate a strategy. Other considerations, such as the differing levels of expertise between the client and the contractor, the inherent risk of failure of the subject of the contract, time constraints, political constraints and other points which have little connection with payment, are all part of the strategic problem. It is therefore essential that the specifier knows the strategic context and, by implication, the terms of payment in order to place the responsibilities correctly. Figure 1.1 shows the relative properties of the different methods of payment.

This model is a common way of depicting the relative qualities of the different types of contract. It is sometimes shown with an additional arrow along the x-axis showing the level of cost control afforded to the client in inverse proportion to the client's flexibility. However, this is too much of a generalisation, and it is quite possible to exercise better cost control in, say, a target-cost contract and still have changes than it is to exercise cost control in a lump sum contract that is subject to changes. Techniques such as 'value engineering', in which the client and contractor assess and agree modifications as the work progresses, can be applied to help control costs.

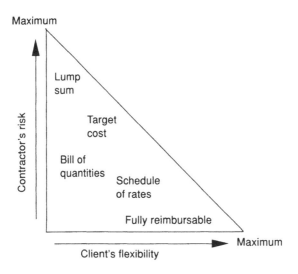

Figure 1.1 The properties of different forms of payment to the contractor

Risks

At one extreme, the risk for the contractor is greatest when the payment is that of a lump sum on completion, i.e. when all the time and technical targets have been met. At the other extreme, it is least risky to the contractor when costs are fully reimbursed during the course of construction up to the point that the technical needs of the client are satisfied.

Of course, there are other risks associated with payment, especially delays, which have serious implications for cash flow or exchange rate calculations. Failure to receive any payment at all is also a risk. However, such risk applies more or less equally to all forms of payment. Complicated financial (as opposed to payment) packages can be assembled to mitigate such risks; but further discussion on this topic is beyond the scope of this book.

Lump sum

Lump sum, in its true form, is, by definition, a single payment for the work. It does not describe the time of payment, nor does it give any clue as to whether the sum is fixed or not (except in the case of pre-payment). The term 'lump sum' is also used to describe fixed-price contracts in which a series of interim fixed-price payments are made against a time or result schedule. Payment of a lump sum in advance means that it must be fixed if the payment is to be for the whole of the work. Lump sum payment at the conclusion of the work may be arrived at in one of two ways:

1 by agreeing the fixed value at the outset;
2 by calculating the values of the work done as time progresses and making a single final payment for work elements individually or totally.

The fixed-price lump sum is clearly what it says, but no indication is given of the timing of payments. Payment may be made as in (1) or by calculation as in (2), in which case it is made any time after completion of the work.

Apart from pre-payment, which is too unusual to be considered further, the nature of the lump sum is that it is usually paid on results. If the time and the technical conditions are not met, then no payment is due. In extreme cases, no payment is due, even if the work is finished and the 'item' is being used by the client, if it is the case that time and technical conditions have not been met.

Fixed profit, reimbursable cost

The logical derivation from a fixed-price lump sum is the fixing of some part of the price. The mechanism of fixing the profit element that the contractor can earn while still reimbursing costs can be useful. In that way, should the contractor over-run the estimate for doing the work, his or her profit, as a percentage of the whole payment, starts to diminish. In extreme cases, it can lead to the contractor working at cost.

Target cost

To differentiate between target cost and fixed price, it is necessary for the client to allow the target to be surrounded by reward and penalty. If the target cost to the client is under-run, that is the work is done for less than the target cost, the client and the contractor share the benefit of the under-run. Conversely, if the target is exceeded, the contractor and the client share the burden of the extra cost. Limits can be placed on the extent of this sharing so that, for example, it is restricted to plus or minus 10% of the target cost. Above and below this range, the contractor will take the risk, as is the case in a lump sum contract.

Target-cost contracts have been shown to be one of the best forms of contract for generating mutual incentives for the client and the contractor, hence leading to teamworking on a project.

Target manhours

Target manhours apply only to contracts in which the client is paying for manhours expended. If manhours are used as the currency of the contract, the cost cannot easily be calculated – unless the manhours are all of equal price. If there is a range of manhour prices, then the contractor should be held to a mix of manhours as well as a pure target. Unless this form of payment is linked to a reward or penalty, it has little value except as a resource measuring device.

Profit sliding scale

If the contractor has been asked to meet a target cost or manhours, then his or her profit can be tied to a sliding scale which is at its maximum at some predetermined under-run, reducing to zero at some point. This is sometimes extended to the point where the contractor starts to repay the client's cost over-run. (This is similar to liquidated damages; see p. 26.)

Measured work, bills of quantities

That method of payment which physically measures the work done and then sets payment according to a list of prices is termed 'measured work payment'; a bill of quantities is such a list.

The measured work form is probably the commonest in civil engineering and building. In this form the designer calculates the quantities in all categories of materials and activities (from excavation to door locks, pipe supports to switches) and lists them out. The contractor sets a price against each item and thereby calculates an overall price. However, in civil engineering and in some building forms of contract (see Joint Contracts Tribunal with approximate quantities, Chapter 2, p. 39), the contractor is paid only for the actual amount of work carried out against each item. If this work exactly equals the quantities in the bills, then the final account will equal the tender. In reality it is complicated by changes, over- and underestimates in the original bills, and circumstances in which the client or contractor considers that the price for an item in the original bill is no longer applicable. Nevertheless, the method is highly flexible and it forms a convenient tool for the specifier to work with, allowing accuracy to the smallest detail.

Fully reimbursable

When the client fully protects the contractor against the latter's costs and adds a profit, the contract becomes 'fully reimbursable'. It is not a particularly subtle form of payment, but it has its uses when neither contractor nor client knows the nature of the work or where it will lead. More common in design than in construction, it is available as a last resort to a client who is unable to persuade any contractor to do the work in any other way. In the case of fully reimbursable payment the onus is on the client to control the contractor and thereby control the costs. From the point of view of the specifier it probably means that (risk considerations aside) there is only a minimum specification with which to work.

Financing

In all forms of payment the contractor will have to finance the work effort until the client pays. The client, on the other hand, must finance payments to the contractor until he or she has the use of the facility under construction. Since the cost of the money is significant, time is an important factor in the overall cost. It is another argument in the set of strategic choices. It should not be assumed that the straight work–invoice–pay chain needs no analysis. The specifier's interest must be in obtaining an economically viable job, and careful specifying can be wasted by incautious economics.

Project financing has always been a key element in the viability of any project and in recent years there has been great interest in what is generally called the private finance initiative (PFI), both in the UK and overseas. Essentially, this system requires a promoting company to raise the capital for the project and then to recover the capital and the financing charges usually over a period, of perhaps 15–20 years, of use. In effect, the ultimate owner of the project is spreading out the payments rather than putting a large amount of capital into a relatively short period of 1–2 years. There are many variations on the theme of private finance.[2]

Incentives

Incentive payments is a curious idea and leads to much debate. There are those who consider that sufficient payment and the prospects of further work are incentive enough, whereas others may feel the need to place money on the table for improvements in contractor performance. The usual criterion of improved performance is an early finish to the work.

All types of bonus payment have value if used correctly. But the specifier may make early finish targets difficult to meet by specifying too strictly. Consider the effect of naming particular suppliers in cases in which a contractor's performance may well be related to the suppliers' workload or even competence. Even specifying a particular independent standard may cut out alternative suppliers. The act of specifying where it is not required, or specifying a particular constraint, can have the effect of making bonus incentives impossible to achieve.

The specifier's primary interest in incentive arrangements should be to ensure that the contractor will not rewrite the specification in order to achieve the bonus targets.

Liquidated damages

Liquidated damages are a contractual remedy for breach of contract: they are commonly in the form of money withheld from the other party or a contractual right to sue for stated sums. Liquidated damages are often written in to engineering contracts to anticipate possible late completion or in the event of unsatisfactory final performance of a process plant.

The whole point of liquidated damages is encapsulated in the word 'liquidated', and it means that the damages are reduced to specific money amounts. In the legal sense damages are compensation for a loss, and liquidated damages are intended to be what the client must receive as compensation for the absence of the item or that which has been promised. Additionally, there is no reason why the contractor should not benefit from liquidated damages for late payment by the client; the usual remedy is bank interest plus a percentage for amounts due under the contract, the interest payments being analogous to liquidated damages.

Two other points arise with the concept of liquidated damages. First, damages are liquidated because both parties agree that in certain circumstances (usually delay or non-performance of some sort) one party pays the damages stated in the contract, in which case for that particular breach no other remedy is available. Second, liquidated damages must be a genuine pre-estimate of the loss suffered by the party to receive them. If they are not a genuine pre-estimate, calculated prior to the award of contract, they will be construed as a penalty and courts will not enforce them. The fact that both parties agreed to them is not relevant.

The specifier's interest here is in the calculation of the loss suffered. In a complex processing stream the specifier will be able to calculate partial outputs corresponding to partial completion. This will enable estimators and others to calculate the money losses. The problem becomes much more complicated if a cost–benefit analysis is carried out as the absence of any benefit (e.g. a footbridge across a road) is not strictly a loss and the specifier can make no contribution to the estimate of loss.

Position of the Engineer

An engineering contract is not a special form of contract; it confers rights and liabilities on those who are parties to it and no one else. This basic tenet is known as privity of contract. However, many engineering contracts provide for the appointment of a person to administer the contract and confer on that person particular powers. The powers arise only out of the agreement of the parties, they do not arise out of the general law and they cannot result out of another contract between different parties. For example, a client may sign a contract with a firm of consulting engineers to obtain designs, specification and supervision of the work. That contract may say that the consulting engineers shall appoint someone (usually a partner or a director) to be 'the Engineer', but the contract between the client and the consulting engineer cannot create obligations on the contractor. The contractor can only accrue contractual obligations from the contract that he or she signs with the client.

The Engineer is a concept; it is purely Anglo-Saxon in origin and, as such, is (or was) unknown on the European continent where other major developments in engineering contracts took place. Because it is an unusual concept (that parties to a contract employ a third party to run the contract for them) the word Engineer (with a capital E) is used often throughout this book. The use of the capital E distinguishes this engineer as an engineer with contractual duties. The same device is not necessary for clients and contractors – they are always parties to a contract. The concept of 'the Engineer' is born out of the historical understanding that clients had no engineering expertise and contractors were mere builders, thus both (but primarily the client as he or she was paying) needed an expert to design and supervise the work. Over the years, the concept was expanded to incorporate the impartiality of the Engineer, although in the early days the Engineer was often the developer and the contractor, too. The zenith of the Engineer's position was reached when he (and 'he' is used deliberately here) was incorporated into standard conditions of contract. This was done by making him into a person whose duties were defined in the contract but one who had no rights under it. In the now standard Conditions of Contract for use in connection with Works of Civil Engineering Construction, published by the Institution of Civil Engineers in various editions since early in this century, the Engineer is presently defined as the person firm or company:

> appointed from time to time by the Employer to act as Engineer for the purposes of the Contract and named in the Appendix to the Form of Tender…

Where the Engineer named in the appendix to the form of tender is not a person then the Engineer shall notify the name of a chartered engineer to the contractor who will assume full responsibilities of the Engineer.

In turn, the Engineer can appoint an Engineer's Representative to:

watch and supervise the construction and completion of the Works.

Throughout the contract there is reference to the Engineer or the Engineer's Representative to act on behalf of the client who is paying for the construction, namely 'the Employer'. For the position of the Engineer to have its intended meaning the contract with the contractor must set out his or her powers. Should the construction contract fail to provide for an Engineer, then the client's agreement with the consulting engineer to provide one is meaningless. The client will have to act on his or her own account. This important fact is sometimes overlooked by specifiers who make erroneous assumptions as to who will inspect the works.

Here, there is immediate conflict because the Engineer is an appointee of the client, not the contractor. In the past the Engineer has maintained his or her position because the client was usually technically ignorant. Nowadays the position of the Engineer is under some threat because many clients – both government and commercial – have engineering departments of their own. Additionally, contractors have become more competent on design issues because of the growth of design and construct contracts.

Nevertheless, the Engineer is often a fact of life, and the specifier needs to be aware if one is to be included in the contract and what his or her duties are to be. The Engineer, then, is a concept, embodying the idea of professional and impartial advice during the currency of the construction contract. The client relies on the Engineer to inspect, measure and oversee the quality of the contractor's work and periodically to certify payments to the contractor. The contractor relies on the Engineer to produce designs and specifications on behalf of the client, and to act as an intermediary between the contractor and the client in matters relating to payment. In addition, both parties rely on the Engineer to ensure that the other party meets their obligations under the contract. Ultimately both sides regard the Engineer as the last resort of impartial judgement before some form of dispute resolution that could ultimately lead to arbitration or litigation. The specifier must find out whether the client is to act through the Engineer or not and draft accordingly. Drafting in such a way as to omit the Engineer is not irrevocable, as delegation of responsibility from the client to the Engineer can easily be made in the conditions of contract.

Chapter 2 looks at some of the current model conditions of contract, including the presence, or not, of the Engineer.

2 Recent developments and standard forms

... and it was just below the rock wall that separates the Second from the Third Ice-Field that Harrer looked back down the staircase: 'Up it I saw the New Era coming at express speed...'

Introduction

In the context of this book, 'recent' means the period since the first edition was written. There have been quite significant changes in the way in which engineering contracts are created, established and administered in this 15–16 year period. It is particularly true of the construction industry, which previously had been going along in much the same way since the century before last. However, nothing in this chapter is meant to contradict or override what has been said in the first chapter. The practices described in the first chapter constitute what is the bedrock of engineering contracts and, as long as the Law of Contract exists, the essential features of how to create an engineering contract and of how to write a specification within that contract will always be in place. Nevertheless, attitudes have changed significantly, and the previous approach, exemplified by compulsory competitive tendering on a project-by-project basis, in which contracting parties often worked to their own agenda, has been seriously challenged. There is a concerted move to change attitudes and significant changes have taken place.

This chapter looks at some of the changes which have taken place in the last decade or so, and what modifications, if any, these moves have had on the way in which the specifier should proceed. The second half of the chapter lists the main standard forms of contract currently in use, together with descriptions of the essential elements of these standard forms.

Industry reviews

Different engineering disciplines go about their contracting business in different ways. Traditionally, building and construction have found it quite difficult to conduct their business in a manner free from dispute. Construction has always been a relatively high-risk business, partly because of the 'one-off' nature of construction projects and partly because of the risks associated with work below ground. Manufacturers of nuts and bolts face far less risk when they perform their contracts because of the controlled conditions under which they carry out their work and because they have performed the contract so many times that they now know exactly how to do it.

Government-sponsored reviews have been carried out on the construction industry from time to time, with varying degrees of effect. There have recently been two government-promoted reports into the state of the construction industry:

- The Latham Report (1994) – *Constructing the Team*;
- The Egan Report (1998) – *Rethinking Construction*.

These have had a significant effect on the way in which construction contracts are conducted. The Latham Report, in particular, appears to have been the catalyst for a change in approach.[1] The report followed from a lengthy process of industry consultation and the outcome was a set of key recommendations, covering such diverse areas as partnering, dispute resolution and the recommendation of the New Engineering Contract as the national standard contract across the construction industry.

The phenomenon of government reports into construction is not new. The Banwell Report, with the rather long-winded title of *Placing and Management of Contracts for Building and Civil Engineering Work*, was published in 1964. It dealt mainly with the way in which contracts were awarded and advocated, including, among other things, the two-envelope system for tender submission. In this system the first envelope containing the technical offer from each tenderer would be opened and analysed before the price contained in the second envelope was considered. However, the impact of the report on the construction industry in the UK was not great.

The Egan Report is the most recent industry review and it goes further than Latham in proposing less formality in contracts to the extent that contracts would not be necessary for construction work. This may be a step too far.

Partnering

This has been, perhaps, the greatest change in the construction industry in the last century and it has taken place over a period of 10–15 years. It is debatable whether this change has been possible because of a more stable economic period or whether it has come about because of a true change in culture in the construction industry. It will be interesting to see if contracting practices in the industry revert to the previous adversarial system if there is a prolonged downturn in the economy. An additional factor has been the privatization of many of the large clients, such as gas and water companies, which has given them the ability to plan their capital programmes over a longer period without fear of money being held back in a government spending moratorium. When these organizations were publicly owned bodies, there was a need for them to be seen to be 'more stringent' in how they awarded their contracts, and compulsory competitive tendering was the only option. Whether this gave best value for money is open to doubt.

Partnering is about doing work in a cooperative way rather than in an adversarial way. It is not about having or not having contracts; it is an attitude of mind. As we have seen, a contract need not be in writing, and whenever there has been a bargain struck and one party does work for another for the payment of money or something in kind there will be a contract. If there is a contract, then the Law of Contract will apply and the courts can be involved in the settlement of any disputes that might arise. Therefore, it is always helpful to have a clearly written record of what was agreed in the deal – the contract document. The specifier's role is the same under partnering as it is under a competitively tendered contract, i.e. to achieve clarity in the contract between the client and the contractor. The danger to all parties is that they start to believe that, because they are working in a cooperative way, there is no need for clarity and certainty in the contract. This is a misconception.

In addition to the contract document, the parties also often prepare and sign a partnering charter, which is simply meant to be a statement of their intent to work collaboratively. The charter is not usually meant to be a legally binding document. and quite often the charter will explicitly state that it is not meant to be legally binding. However, the courts have decided that they will interpret a party's actions in a dispute in view of how that party said it would behave in a partnering charter.

The choice of the word 'partnering' for this collaborative way of working can also lead to some confusion about the strict interpretation of the relationship between the parties. They rarely intend to be partners in the sense of a partnership company, but they may often sail close to that wind if they sign agreements, share offices and share profits. The parties have to be careful with the words they choose and the actions they take. Partnering is now used by many large organizations, particularly those which have a continuing programme of work. The benefits are that the two parties can work together to develop the best overall solutions to deliver the client's projects by looking at the programme as a whole rather than by looking at individual projects as they come along. Partnering contractors are able to take part in and assist the client in many areas such as early project risk assessment and supplier liaison.

A concern of many clients is whether or not they are achieving best value for money when partnering is compared with competitively tendered projects. Generally, the original partnering contract will be arrived at through pre-qualification from a list of contractors and subsequent competitive tenders, based on a partnering agreement for perhaps 3 or 4 years. The original partnering contract will contain the payment mechanisms which will usually include rates and fees and possibly some incentive systems. However, the client can gain some confidence in the value for money achieved by the partnering arrangement by leaving a small portion of the programme of work to competitive tendering. Comparative valuation is one of the most difficult aspects of engineering contracts and, because estimating allows a margin of error, the true value of working in any particular way is never easy to prove or disprove. One way in which the client can truly verify that the partnering arrangement is providing value for money is to allocate between 10% and 20% of the total spending on capital projects to alternative contracting arrangements, namely competitive tendering.

Supply-chain management

Supply-chain management is a popular piece of current jargon, but its saving grace as a piece of jargon is that it is quite descriptive. In practical terms, it is concerned with the way in which the client might influence how goods and materials, at a quite detailed level, are supplied for projects. Manufacturing industry has always practised supply-chain management as a matter of survival. The consequences of not doing so for a vehicle production line would be very serious delays and losses. Informal supply-chain management has been practised for many years on engineering, construction and building contracts through such mechanisms as nominated suppliers and the free issue of goods and equipment by the client. However, these practices fell into disrepute in some areas, mainly because of the adversarial attitudes often displayed in main contracts. The late delivery of free-issue materials or the default of a subcontractor who was nominated by the client were common causes for contractors' claims. With the development of partnering in contracts and in programmes of work, management of the supply chain has been made easier for the client. Many large clients feel that they can influence suppliers to provide better prices and better service in return for a special trading arrangement such as a framework agreement.

Framework agreements

Framework agreements are a useful tool in supply-chain management. They take the form of periodic agreements between a client and the suppliers of materials and equipment that are commonly used by the client. The parties agree a set of trading terms and prices, against which supply contracts will be placed by the client or possibly by contractors working on projects for that client. The individual contracts can be formed by a simple exchange of letters which refer to the framework terms. The framework agreement itself lacks certainty because there is no guarantee on the amount of the supply. However, the separate contracts do have certainty.

There are many pros and cons around the effectiveness of framework agreements. They are most commonly put forward as a means of buying at the lowest possible price because of the client's 'buying power'. However, it is not unusual for the same items to be bought at a lower cost under different arrangements, though this is often because of some different circumstances and possibly because of some subtle difference in the specification. The benefits of framework agreements come from better prices, but perhaps more importantly from:

- standardization;
- preferential service.

Standardization

Standardization of equipment has many potential benefits, not only in the specifying and construction phases, but also in the subsequent operating period. It is far easier for a client's maintenance team to service perhaps six different makes of pumps than it is to service sixty different makes of pumps. Framework agreements can force this degree of standardization in contracts in which good intentions towards standardization would previously have been thwarted by preferential engineering. There will be times when the framework agreement supplier of equipment will not be appropriate to a particular application, and in these circumstances a waiver will be agreed for an alternative bespoke supply, but this will be the exception rather than the rule.

Preferential service

Preferential service may be a misnomer, but there are undoubtedly service benefits to the client in having a more formal trading arrangement with key suppliers. Obviously, this relationship will become strained if the framework agreement is in name only and the volume of orders does not materialize but, provided that the framework agreement is used as intended, there will be service benefits to be had from such an arrangement.

Framework agreements tend to work best with commodities rather than with the supply of complex engineered products. The latter often require some design input to tailor them for the particular project, and this will inevitably lead to uncertainty around the final price and to a loss of the main advantages of a framework agreement.

The Internet in engineering contracts

The arrival of the Internet has brought many potential advantages to the establishment of engineering contracts. Essentially, the effect of the Internet on engineering contracts comes under three headings:

- e-intelligence;

- e-selling;
- e-buying (or e-procurement).

The last two categories are opposite sides of the same coin and they are often referred to collectively as e-commerce.

e-intelligence

The availability of information on the Internet is truly astounding. Specifiers can now access a whole reference library of information through their own desktop computer. The sheer volume of information that is available can be a drawback, but the benefits far outweigh this. British Standards can be accessed and downloaded, by subscription, as can information about the American Pipelines Institute and virtually any other body of significant engineering interest.

e-selling

The Internet is also a powerful medium for advertising, and many companies, including engineering companies, advertise and sell their products through the Internet. The specifier's interest in e-selling is in the wide availability of goods and materials that may be of use in the engineering contract. The specifier has to be confident in the standard to which that article is manufactured. Kite marks and CE marks[2] are invaluable guides, but the specifier must beware of any unspecified areas of detail; e-selling can offer suppliers a tremendously powerful medium to demonstrate and advertise their products. For example, a brick manufacturer can display the appearance of any of its range of bricks in combination with a wide range of different coloured mortars. The specifier can view a panel of a particular brick and then select the mortar colour from a large pallet of colours to see the overall effect on the computer screen. The total number of combinations can be in the thousands. The resultant display may not be as true to life as a trial panel, but it will be a very effective way for the specifier to hone down his or her selection.

e-buying

e-buying is, of course, the reverse of e-selling, but it does also describe a particular method of buying. Many clients have explored the possibility of 'buying' engineering contractors by using the Internet. The way this is done is through a 'reverse auction'. In a normal auction, bidders try to bid the highest price to buy a piece of art or a used car, depending on what is being auctioned. In a reverse auction, the seller is selling the opportunity for the buyer to win the contract. Therefore, the potential contractors try to offer the lowest price for the contract.

The whole process of the reverse auction assumes that bidders are qualified to bid for the work and that they have been through an appropriate vetting or pre-qualification procedure before being given the access rights to take part in the auction. However, there are two areas of concern in this process. The first is that engineering contracts are complex contracts which need to be studied and assessed over a period of time. The idea of bidding on the Internet for a complex design and construct contract for a process plant is frankly foolish, unless it is the final step in a detailed tendering process that has run for several weeks. Even then, a bidding process might introduce last-minute flaws in the contract that would be better avoided. However, a supply-and-erect contract for two storage tanks on an existing base slab might be a suitable candidate for this approach. The second area of concern is that the reverse auction is a process that might not necessarily yield the best offer, because the potential contractors can see what the alternative prices are and so they only have to offer a slightly lower price. Competitive tendering via sealed offers, on the other hand, always requires each tenderer to make his or her best offer. Moreover, if the concept of reverse auction were a sound way of placing engineering contracts, it would, no doubt, have been tried and become commonplace long before the Internet came into being. After all, all that is required to conduct an auction is an auction room.

Directives from the European Union

Our membership of the EU affects various aspects of our working and domestic lives, and the specifier must be aware of how directives from the EU can affect engineering specifications. First, it is useful to be aware of the structure of the legislative bodies in the EU. The three key bodies are:

- the European Parliament;
- the Council of the European Union (Council of Europe for short);
- the European Commission (EC).

The Members of the European Parliament (MEPs) are elected every 5 years by the people of Europe – some 374 million of them have the vote. The Parliament has legislative powers which it shares with the main EU decision-making body – the Council of Europe. The Council is made up of representatives of the governments of the member states at ministerial level. The European Commission proposes EU directives. The Commission consists of appointees who are put forward by the countries within the EU and ratified by the European Parliament. The Commission also has to enforce legislation once Parliament and the Council have approved it. The member countries of the EU have a duty to adopt the approved EU legislation so that it becomes part of the law of that country.

An example of an EU directive that has affected some engineering contracts is the Council directive concerning the procedures for the award of supply, works and services contracts in the water, energy, transport and telecommunications sectors. This directive has been adopted into UK law under a statutory instrument – The Utilities Contracts Regulations 1996.[3] Essentially, the regulations require public organizations and private companies, working as clients in these sectors, to advertise projects across Europe. There are mechanisms by which restricted procedures can be put in place, but the main objective of the regulations is to open up the tendering process and to provide equality of opportunity for contractors and suppliers throughout the EU.

Industry standard forms

The construction industry employs standard forms of contract in two specific situations: the client's relationships with contractors and with consultants. As engineers become more directly involved in construction contracts than in consultants' contracts, the former are better known; in addition, the world of consultants – coming as it did from a base of partnerships – tends to regard consultants' contracts as being not for general consumption. There are standard forms for consultancy services, and the Association of Consulting Engineers (ACE) produces a comprehensive set of documents to cover the range of services on offer.

Standard forms exist because they have a purpose or area of responsibility, and because they *are* standard. In many cases, both ideas seem to be forgotten – inapplicable standards are used without proper thought and redrafted with little understanding. There is a tendency for clients and their advisors to modify industry standard forms by the addition of a multitude of special conditions, most of which are written to shift the contractual advantage towards the client. Special conditions of contract are necessary for project-specific issues, and some standard forms provide a structure and guidance notes to assist with the writing of such special conditions of contract. However, if the client seeks to modify many of the general conditions of contract, the advantages of using a standard form will be lost. The need for standards arises because contractors and clients value certainty in their relationships. In the legal sense, certainty arises out of court decisions on particular standards. As it may take 5 or more years for definitive case law to be of direct use, the development of new standards is difficult. However, there has been a recent surge in the publication of new editions of standard forms by most of the issuing bodies.

No doubt the selection of a standard form of contract will be made by management, but the specifier should have some awareness of the standard forms in order to appreciate the implication of choices made. The standard form should, at least, reflect the client's and the contractor's risks, consistent with the specification. To concentrate on important matters as far as the specifier is concerned, various standard forms appear in the following pages; in each case there are notes on the following items:

1 the client–contractor relationship and the position of the Engineer (if he or she exists);
2 the calculation of the contract price;
3 variations to the specification.

There are, of course, many other key aspects to each model (or standard) form, such as how disputes are resolved and what the obligations of the parties are, but a discussion on all the aspects of each form would require a separate book. Furthermore, discussion is restricted to industry forms and does not address the myriad local authority forms or variations to the standards, or UK and foreign government forms, or those used by public utilities. The illustration of some UK industry forms, then, should be sufficient to introduce specifiers to the standard forms.

ICE conditions of contract

The Conditions of Contract and Forms of Tender, Agreement and Bond, for use in connection with Works of Civil Engineering Construction, otherwise known as the ICE Conditions, are sponsored by the Institution of Civil Engineers, the Association of Consulting Engineers and the Civil Engineering Contractors Association; the Conditions are now in their seventh edition. A widely used standard form, from the first they have established the position of the Engineer as the client's representative throughout the onshore civil engineering industry.

The scope of the contractor's responsibilities is to: 'construct, complete and maintain the Works' and the contractor is only meant to design any part of the permanent works if the contract expressly states so; the ICE publish a separate set of standard conditions to cover design and construct contracts. The client is known as the Employer, and delegation to the Engineer is absolute where it exists, i.e. where the Employer has delegated his rights to the Engineer in the terms of the contract, he cannot reclaim them without the consent of the contractor. The contract price is defined but not in money terms; it is the 'sum to be ascertained and paid… in accordance with the Contract'.

The specification means 'the specification referred to in the Form of Tender and any modification thereof or addition thereto as may from time to time be furnished or approved in writing by the Engineer'. A similar definition applies to drawings. Taken in conjunction with the power of the Engineer to order variations which 'may include additions omissions substitutions alterations changes in quality form character kind position dimension level or line…', the power to make changes seems wide enough. The problem is to prevent changes to the specification (through the Engineer) totally disrupting the contractor's work. Variations have a price, calculation of which is allowed for in the contract.

The New Engineering Contract

The Institution of Civil Engineers also publishes the New Engineering Contract (NEC). Since it was first published in 1993, the NEC family of documents has been added to with the addition of a 'short contract' for relatively straightforward work, a professional services contract and an adjudicators' contract. The standard forms that formed the original NEC are now collectively known as the Engineering and Construction Contract (ECC).

The standard form consists of a separately bound set of core clauses that is used together with one of several optional clauses, depending on what type of contract is chosen. The main options are:

1   priced contract with activity schedule;
2   priced contract with bill of quantities;
3   target contract with activity schedule;
4   target contract with bill of quantities;
5   management contract.

There is also a standard form of subcontract in the suite.

Each option has additional clauses to the core clauses, and these are shown in bold type to aid identification. Secondary option clauses are also written in to each main option document to provide for matters such as retention and performance bonds, if the client requires them. The parties to the contract are the Employer and the Contractor, but the term 'Engineer' is not used and the functions normally undertaken by the Engineer are carried out by 'the Project Manager' and 'the Supervisor'. The contract price is calculated in different ways depending on the choice of main option. Variations are provided for under the core clauses in which they are known as compensation events. Here, the NEC is perhaps more detailed in what is and what is not a compensation event (variation) than other standard forms. There are strict time constraints for submissions concerning compensation events which can be quite demanding of the parties. The family of documents also comes with a booklet of flow charts that give detailed instructions and guidance on how to proceed under the contract clauses in various situations.

The NEC represents a radical leap in standard forms of contract, mainly because of the style of the language that is used. All the other standard forms of contract tend to direct in their clauses as to who does what and who should do what – and this is stated in varying degrees of legalese or plain English. The NEC is written in a more descriptive style, for example 'The Contractor obeys an instruction which the Project Manager or the Supervisor gives him and which is in accordance with this contract'. Needless to say, the absence of the direction 'shall' is a cause of some concern in the legal profession. However, the form is now widely used and has many devotees.

The NEC was written specifically to encourage better site management and a partnering approach, and it was recommended in the Latham Report as the standard form of contract across the whole area of engineering and construction work.

Standard Form of Building Contract

The Standard Form of Building Contract, 1998 Edition, produced by the Joint Contracts Tribunal, commonly known as the JCT or RIBA (it is published by the Royal Institute of British Architects), is used mainly on building work. The JCT is made up of the Royal Institute of Chartered Surveyors (RICS), RIBA, local authorities, the Association of Consulting Engineers, and contractors' associations. Notwithstanding its building bias, because of the blurred dividing line between building and civil engineering, it is useful to be aware of the provisions of this standard form. The contractor is required to 'carry out and complete the Works'. The word 'construct' is not used and the definition of 'Works' is not prescribed, but is left entirely to the parties. However, it seems implicit that 'Works' is an entity, not an activity.

The JCT form exists in six variants, three for local authorities and three for private clients, thus:

- with quantities;
- without quantities;
- with approximate quantities.

There are also shorter forms depending on the complexity and value of the project. The intermediate form of contract (IFC 98) and the minor works form of contract (MW 98) are intended for use on projects with approximate values of up to £400,000 and £100,000 respectively.

The client is known as the Employer, and he or she delegates duties to the Architect. The powers of the Architect are absolute where they exist and cannot be unilaterally reclaimed by the Employer. The price of the Works is known as the 'Ascertained Final Sum', though its ascertaining varies slightly in the variations of the form. The Architect has powers to issue variations which can add, omit, substitute, alter the kind of, or standard of, any of the materials or goods to be used in the Works as well as varying the Works. These powers are wide enough to disrupt the contractor's work by changes to the specification, thereby incurring extra costs, the calculation of which is allowed for in the contract.

The model forms – electrical and mechanical work

A series of model forms is published under the title *The Model Form of General Conditions of Contract (including Forms of Agreement and*

Guarantee) by the Institution of Electrical Engineers, in conjunction with the Institution of Mechanical Engineers. The model forms exist in three versions:

1   MF/1 Home or Overseas Contracts with Erection;
2   MF/2 Home or Overseas Contracts for the Supply of Electrical, Electronic or Mechanical Plant
3   MF/3 Home (without erection).

The Association of Consulting Engineers endorses forms MF/1 and MF/2, but it has no involvement with model form MF/3, which does not include provisions for the appointment of an Engineer. Of the three versions, model form MF/1 is the most widely used, and the contract is between the client as 'Purchaser' and the Contractor. There is provision for an Engineer if one has been nominated by the Purchaser; if one has not been nominated, the Purchaser becomes the Engineer.

The Contractor executes the 'Works' and the 'Works' means 'all plant provided and work to be done by the Contractor under the Contract', but the model form relies on the specification for definition of what 'work' the Contractor shall carry out. Therefore, the specifier may need contractual advice on the 'scope of work' description in the specification. Payment is not referred to, as such, but the contract price is the 'sum named in the Contract' and there is a 'Contract Value' which is 'such part of the Contract Price, adjusted to give effect to such additions or deductions as are provided for in the Contract,... as is properly apportionable to the Plant or work in question...'. Only the Engineer has the power to order variations before the whole of the works have been taken over. There is a restriction on the amount of variation to the works that the Engineer can order without the Contractor's written consent, which is stated as that which 'involve[s] a net addition to or deduction from the Contract Price of more than 15 per cent thereof'. Thus, to some extent, changes to the specification can be controlled by the Contractor.

The four FIDIC standard forms

In 1999, FIDIC brought out four new model forms of contract. The abbreviation FIDIC stands for Fédération International des Ingénieur-Conseils, that is the International Federation of Consulting Engineers. FIDIC was founded in 1913 in France, Belgium and Switzerland, and today it has its headquarters in Switzerland. The English feel of the documents is quite surprising, but that is because of the history of the use of the model forms over the years.

As with many model forms of contract, the FIDIC forms have come to be identified by the colour of their cover – the 'orange book', the 'red book', etc. However, FIDIC would like the latest versions to be known by what they cover rather than by their covers. Also, the colours of the covers on the latest editions do not coincide with the colours of the previous editions. The four new FIDIC forms are:

1 Conditions of Contract for EPC Turnkey Projects;
2 Conditions of Contract for Plant and Design–Build Projects;
3 Conditions of Contract for Construction Projects;
4 Short Form of Contract.

The EPC turnkey contract

The parties to the contract are the client, known as the Employer, and the Contractor. There is no Engineer in this form, which is intended for use on large complex projects that probably involve extensive commissioning, and in which the client wishes to transfer all the construction risk to the contractor. The name 'Turnkey' derives from the idea that all the client (or more likely the separate operating contractor) has to do is come along and turn a key to start using the installation.

The Employer's Representative represents the Employer on site, and it is clear in the conditions that the Employer's Representative is acting on behalf of the Employer and that he or she is not impartial. The contract price is normally a lump sum price, subject to any contractual adjustments. The Employer can order variations, either as a result of his or her own requirements or as a result of some proposal from the Contractor that is beneficial to the employer. As with all design and construct forms of contract, the specifier has to be very precise in defining what is the specification and the scope of the works that are in the contract. The exact description can become clouded in the exchange of enquiry documents and tenderers' proposals. If this is the case, it can be very difficult to agree on what is and what is not a variation in this form of contract. The conditions do address the question about the hierarchy of the various contract documents, but the respective merits of the specification and the contractors' design proposals at the time of tender can still be a difficult area in design and construct forms of contract.

Plant and design–build contract

The parties are the client, known as the 'Employer', and the 'Contractor'. The form is a design and construct form and it is intended for projects for the provision of mechanical and electrical plant and for the design and construction of engineering works. In this case there is an Engineer who has duties and authority derived from the contract. However, the conditions clearly state that the Engineer is deemed to be acting for the Employer. The contract price is a lump sum price and the contract allows for an advance payment mechanism that is recouped over subsequent interim payments.

Variations and other price adjustments are allowed for, together with a calculation for price adjustment caused by fluctuating costs.

Construction contract

The parties and the Engineer are the same in this contract form as they are in the plant and design–build form of contract. Again, the Engineer is deemed to act for the Employer. The contract price is 'agreed and determined' under the contract in accordance with the contract method of measurement, together with agreed rates and prices. The conditions also clearly state the rules under which contract rates can be varied, one of which is by variations which can be ordered by the Engineer or which arise from a value engineering proposal put forward by the Contractor.

The short form of contract

This is intended mainly for client-designed projects of relatively small capital value but with provision for contractor design. The parties are the Employer and the Contractor and there is no Engineer, but the Employer must appoint an authorized person and the Employer may appoint a firm or individual to act as the Employer's Representative with any powers that the Employer cares to delegate. The contract price is 'valued', as provided for in the contract, and it can be valued by any means from cost reimbursable to lump sum price. The details are recorded in the appendix to the contract agreement. Variations are allowed for, and the contract states that 'the Employer may instruct Variations'. The Employer has quite wide powers in instructing and valuing variations.

Model form for process plants

The Model Form of Conditions of Contract for Process Plants is published by the Institution of Chemical Engineers (IChemE) and comes in two versions – one suitable for lump sum contracts in the UK (the 'Red Book'), now in its third edition (1995), and the other suitable for reimbursable contracts (the 'Green Book'), now in its second edition (1992). The parties to the contract are the Contractor and the Purchaser, and there is provision for a Project Manager. The Project Manager is clearly the servant of the Purchaser and 'any obligation stated under the Contract to be an obligation of the Project Manager shall be deemed to be an obligation of the Purchaser'.

The lump sum version defines the 'Contract Price' as the 'sum named as such in the Form of Agreement', but adjustments to the contract price 'shall be determined by valuation of Variations', the Project Manager having the power to order in writing a 'Variation' – i.e. 'any alteration in the Plant or in the type of or extent of the Contractor's work, with special reference to the Specification... being an amendment, omission or addition'.

The reimbursable version defines the 'Contract Price' as the 'total sums payable by the Purchaser', and the sums are then calculated in accordance with schedules attached to the conditions. There are detailed notes on the preparation of the schedules, setting out what should be priced and how items are normally split for pricing purposes, including calculation of variations. The valuation of variations is straightforward in a cost reimbursable form of contract, because variations are valued in exactly the same way as the rest of the contract works. The main difficulty is the interpretation of what will or will not be allowed as a reimbursable cost, such as a head office charge. The distinction is not always clear when the project is under way. The reimbursable form has been quite widely adapted for use as a target-cost contract.

The model form for process plant is unlike any of the other engineering contracts in that it deals in detail with matters that the others gloss over such as process performance testing. Each form comes complete with extensive notes which are worth study in themselves.

Other contract types

Because a contract is a bargain, it can take any form that the parties desire. Most of the industry standards are concerned with supervision of the construction process by others on behalf of the client, but there can be any number of permutations of the design–supervise–construct arrangements, including:

- design and construct;
- design and manage construction (with or without the client);
- construct and commission;
- construct and operate.

The distinction between management and supervision is a nebulous one, but it serves to illustrate the difference in approach of the client who may be an expert and wants to manage the construction in association with a designer and that of the client who is not an expert and delegates supervision. But even an expert client may delegate the management if he or she is temporarily short of the resources to manage. In any case, the so-called management contract is usually one of the above mixtures, and the name itself is often little more than a marketing label as it rarely describes the depth of management or the actual contractual responsibilities.

Special difficulties arise over client control of the specification in contracts in which the client has a place in the management of construction. This is because the designer will have had control of the design process and then the client becomes involved in actually putting the design into practice. Involvement of the client at this stage has to be carefully controlled if costs are to be contained.

Sometimes a contract is put together which includes the Engineer but in which the documents do not use a recognized industry standard, or are only loosely related to one. Although there may be a need for this role – really a project manager – it will not be couched in the form that the industry may have come to regard as the norm and can lead to confusion. The use of the concept of the Engineer, in the sense of an impartial manager/umpire for the contract, without the relevant standard form is best avoided.

3   Presenting specifications

You see it's like a portmanteau – there are two meanings packed up into one word.

Before drafting

Drafting a specification is not a launch into the unknown like writing a novel, when the characters can develop as the novel progresses. Before drafting commences, the specifier must know the technical context in which he or she is to write, which will comprise the client, the client's requirements and the overall technical design to meet those requirements. Specifiers need to have knowledge of the feasibility stages of the project within which they are to work in order to appreciate the constraints governing it.

The specifications present not only documents (such as drawings), but also other people's ideas which need to be amalgamated. In writing a specification for the disposal of sewage from a building, it would be unfortunate to specify a urinal if the client had expected a gentlemen's washroom.

The form of contract

The form of contract within which the specification will sit will have been determined by the commercial considerations of the client and the client's advisers. The most common standard forms are described in Chapter 2, and they fall into the following general categories:

- construct only;
- design and construct;
- design and construct, including mechanical and electrical (M&E) and process plant.

Within these general categories there are many variations depending on how much design the contractor will carry out. Confusion can occur in some forms because they include 'schedules' as part of their standard structure. The IChemE forms have schedules that deal with specific matters, such as performance tests and documents for approval, but also with more general matters such as the description of the 'Works'. Such matters would be covered in the appendix to the form of agreement, and in the specification in those standard forms that do not contain schedules. However, if the standard contract form requires that schedules are included in the contract, it is important that duplication is avoided. In such forms duplication can easily occur in the general specification items. Therefore, the best way to treat the matter is to regard the schedules as a separating out of the important parts of the specification and to write the schedules using the same rules that would be applied to the specification.

Technical and general specifications

In most engineering contracts the specification is divided between two areas: the general requirements of the work and its control by the client, and the technical details. Although the technical details are invariably referred to as the technical specification, the more general areas of description and management control are variously referred to as the general specification, scope of work, Engineer's requirements, coordination procedure, management section, and so on.

The general requirements (whatever name they come under) usually cover a brief description of the work, together with sections on (as appropriate) reporting to the client, issuing of instructions, responsibilities of individual posts in client's and contractor's organizations, control of costs, planning, offices, meetings and minutes, transport, shipping, procurement, placing subcontracts, use of computer systems, quality assurance, stores management, import regulations, employment of personnel, and all the other attendant requirements.

The technical specification, in consequence, should avoid all matters that are not purely technical. It makes for a very confusing list of general obligations if the specifier scatters the general requirements like confetti over the technical details. If there are comments to be made, for example on programming and planning, then the specifier should collect these together into a single section, so that the user can appreciate the planning requirements at all levels.

In this and subsequent chapters there is the assumption that the specifier is going to be involved in drafting both the technical and the general specifications. In fact, it is often a natural progression for specifiers to start on technical details and then to develop their careers into those management-related activities encompassed by the general specification.

Presentation

The specifier, then, is about to commence exercising choice. Apart from the appropriateness of what is to be written, which will be absorbed through reading the client's requirements in a design brief or concept outline, a decision on how to present it must be made. Presentation requires choice among three areas:

1   style of communication;
2   type of specification;
3   source of specification;

Style – method or result?

There are basically two different styles in which specifications can be presented and these are:

1   the method specification;
2   the results specification.

The specifier must find out from the client exactly what the priorities are, in terms of how he or she wishes to achieve the end-product, and use the appropriate style or combination of styles accordingly. At one extreme, the specification can be a simple statement of result, and this is often of greatest importance to a producer with a product to sell. At the other extreme, in which aesthetics are of most importance, detailed method specifications may have to be written into the contract. Method should not be confused with detail. It is possible to specify a great deal of detail about the final quality of the works and still have what is a results specification. Indeed, style is probably the single most important matter that has to be considered. In the absence of style, the specification quickly degenerates into a mish-mash of instructions and expectations. However, this is not to say that style is not influenced by the type and source of specification, as will be seen further on in this chapter, as one style may be unavailable to the specifier because of the choice of type or source.

An important point is that, although a mixture of styles is acceptable in looking at the specification as a whole, it is not acceptable within a single operation. Consider the following specification for road construction concerning the California Bearing Ratio (CBR) method of defining ground bearing capacity:

> The Contractor shall achieve 95% CBR with 15 passes of a five-ton roller.

This is a good example of confusion of styles. On the one hand, the result is specified – 95% CBR – and on the other hand, the method – fifteen passes. Quite clearly, the result may be unachievable, especially as the roller weight is specified as well. In this case, the specifier must choose how he or she is going to specify the client's requirements, either a road to the required CBR standard (the results specification) or the number of passes with the specified roller (the method specification). It would seem that in the example there is an obvious preference for the results specification, because the result is of utmost importance to the client. However, such a choice depends on other constraints, for instance the proposed plant fleet may only have five-ton rollers, or time for rolling may be limited, or the designer may consider that over-rolling may damage the subgrade. Even in such a simple specification, the choice of style is important. It does not automatically flow from the choice of specification itself.

A mixture of styles will be inevitable in any specification because of the varying control the specifier may wish to see exerted on the contractor. However, in a single operation, such as the road compaction example, there must not be a mixture of styles. Every element of the job will demand its own style, and the specifier's task is to choose the correct one.

Method specification

The method specification consists of instructions to the reader on how to achieve a particular result, the actual result appearing only as an afterthought to the method description. Consider the following example:

> The Contractor shall make provision for the pressure relief and venting for gas tanks by the insertion of a valve at the top of each tank.

At first sight, this may seem a waste of words. The specification might well have stopped after the word 'tanks', because everyone knows that gas under pressure with a liquid can escape only from the top of the container. However, the contractor could have set a side-valve near the base of the tank with an internal pipe going to the top on the inside of the tank. Such a solution still vents gas from the top (as it must) – and to a lump sum contractor would be the more attractive solution because of the savings of walkways and ladders on the top of the tanks. The specifier considered the internal venting to be intrinsically unsafe because of the possible blockage of the internal pipe, therefore justifying the use of the method specification. However, it is not so subtle to write:

> The Contractor shall construct each wall by laying courses of brick starting from the bottom.

This illustrates the danger of a decline into the absurd.

The advantages of the method specification are mainly for the client. A client who wants to maintain control over all parts of the construction process would be sensible to specify each stage with some precision. Of course, the reason why the specifier chooses such a style is dependent on the type (see p. 54) and source (see p. 59) of specification. The greatest advantage of a method specification is for the expert client employing a contractor as the builder. Unfortunately, some clients, though not expert, do use method specifications and thereby restrict the contractor's choice of optimum solutions – to no conceivable advantage for anyone. Contractors are employed for their expert practical skills in construction or M&E installation or whatever discipline they come from; the idea of employing an expert and then instructing them how to do their work is anathema to most specifiers.

The extreme use of the method specification is in contracts in which the client holds patents or other proprietary rights over a method and is instructing the contractor to use that method. In all method specifications the contractor's liability in contractual terms is satisfied when the method has been correctly applied – whether or not any implied result has been achieved. In fact, in a case in which certain bricks had been specified for a manhole and the manhole, though correctly constructed, had leaked, the courts decided that the bricks met the specification and therefore the contractor was not liable.

The risk of failing to achieve the result intended is the most obvious disadvantage of the method specification. The client cannot easily specify method and result, but might have some chance of doing so by approving methods. However, the risk of failure of the method is mitigated if the contractor (after the work has started) or the tenderers are allowed to comment on the specified methods. In this way, the specifier can leave some options open by allowing the contractors to bring their expertise forward. However, a specifier who is intent on specifying method must conscientiously receive the representations on the method and then clearly confirm or alter the original specification. It is not unusual for some sort of understanding to be reached on divergence of method only to find (too late) that it was really a misunderstanding.

If the successful tenderer is insistent on substituting his or her method for the client's, the specifier must make sure that the contract documents clearly show the method to be the contractor's. In simple cases, sufficient protection should be afforded by wording such as:

> The Contractor warrants that the method described in Section XX will achieve the intended result and additionally agrees to revert to the method in Section YY at no cost to the Client if the Contractor's method fails.

One of the most difficult areas of the method specification is the manner in which it first appears out of the tendering process. Many tender documents ask the tenderers to supply their method of working, either for some part of the works or for the job as a whole. The client has only two reasons for making such a request: to ascertain if the contractor understood the tender and to evaluate the method; or to incorporate the method of working into the contract as part of a method specification. Both reasons are valid. The former is a sort of pre-qualification; the latter is a way of obtaining benefit from the tenderer's expertise. However, it becomes a perversion of the tendering process if the client requests the method for the pre-qualification process and then incorporates it into the contract. In such a case it is almost certain that neither the tenderer nor the client will have drafted it for that purpose, and its inclusion may lead to contradictions in the contract.

Should the client require that the contractor's method of working be included in the contract, the specifier must check that the conditions of contract correctly apportion liabilities – that is do not absolve the contractor from producing the desired result – and that the method itself is properly drafted as a contractual document. There are probably two ways of drafting the working method:

1   as a base plan in overall concept form;
2   as a detailed description complete with approvals.

In addition to the main contract and whatever provisions it makes for the method of working, in situations in which the client nominates subcontractors or suppliers it is very often the case that such nominations come from a desire to settle on a particular method in which the subcontractor or supplier is expert. This can occur in contracts in which no methods are specified otherwise, and the specifier must ensure that the nomination does not imply a method specification that was not intended. Of course, in all choices of contractor there is an implied method of working, but this need not be explicitly stated and the specifier can rely on the original (or other) specification.

Another version of this situation occurs when the client gives staged approval to the contractor's methods but has relied originally on a method specification. Quality assurance systems seem to have made this approach more common than was previously the case. Whereas quality control was concerned with the outcome of any process more or less independently of the process itself, quality assurance focuses on the process. Quality assurance requires that the process be properly operated, based on the theory that the proper operation of a sound process will produce the correct results. The outcome of quality assurance is that the client becomes involved in approving procedural descriptions of methods, even though the client may have specified only results. However, quality assurance systems have not convinced many clients to give up quality control at the results stage. For example, in the offshore industry the client not only specifies welds in terms of dimensional, material and strength details, but also specifies that the procedural testing of welders be subject to the client's approval. The existence of some quality control stage, such as the cube strength sampling of concrete or the proof testing of a bored pile, effectively becomes a part of the quality assurance procedure. This is a form of method specification which is not intended to over-ride the results criteria. The specifier must take into account the quality of the workforce when writing the specification, even though the specifier is not really writing a method specification.

When the client is providing the equipment or plant to carry out the work (as in much aid programme work), the specifier is put in a very difficult position. It is sometimes the case that the nature and quantity of the plant to be provided is not known; for example, the specifier, together with the designers, may be working on the premise that reinforced concrete will be used, whereas in fact the client cannot make provision for any handling facilities for the steel reinforcement. Roads are particularly vulnerable to this sort of 'specification by plant availability'. If a specifier writes a specification for a bitumen-surfaced road and there is no bitumen plant, the contractor will obviously leave out the bitumen coat. However, the subgrades and sub-bases for bitumen and gravel roads are quite different, and the end-result will be an inferior road. The specifier must be aware of any form of limitations on plant or equipment set by the client so that the method specification includes the correct assumptions.

Method specification need not concentrate on how to carry out construction or assembly; it can also extend to management of a design or construction process. The following is a method specification aimed at controlling a management technique:

> The Contractor shall plan the Work using network analysis in activity-on-the-arrow form, providing three network levels.

Similarly, the following specifies a method (the client's) of coding costs that the client has to use:

> The Contractor shall use the Client's Account Codes when allocating costs of the Work to individual design and management tasks.

Similarly again, if a procurement contractor has to use the client's conditions of purchase and bid procedures when operating on the client's behalf, then that contractor is working to a method specification.

In all of the above cases the client has decided that his or her methods are the most suitable form of management. This process can go even further into details, for instance specifying formulae to be used in design calculations.

There is another way of specifying method – by stealth. The situation in which a contractor is not working to a method specification, but nevertheless has to allow the client to approve the methods of work, can be misused by the client to specify method by the back door. When a contractor meets continual refusal to adopt his or her own methods, that contractor will end up adopting the client's methods; but the sensible contractor will issue and create appropriate and contemporary written records in an attempt to gain absolution from liability for their failure. This is not the same as negotiating method at the time of tendering, because, by this later stage, the work will probably be well advanced and the contractor's programme could meet costly delays if agreement is not reached.

The specifier, in adopting the method style, has to balance guidance against interference in order to obtain the best results.

Results specification

The specifier writes a results specification simply by setting down for the contractor the criteria that the latter has to meet at the end of an operation, or even, in extreme cases, the whole job. The specifier makes no reference to the method of achieving the result, except perhaps by making reference to intermediate results. The results specification is also a performance specification, that is the performance of any particular item or job is the 'result'. Performance is also used to describe the result, which may be at an interim stage in a contract, such as a proof test on a 70 kN pile, or at the final stage of a contract, in which several performance measures may be specified in a schedule. The measures may include requirements for:

- power consumption;
- chemical consumption;
- noise levels;
- quality of final outputs from the works.

The terms 'result' and 'performance' are synonymous. The results specification is the normal way in which a non-technical client will set out his or her requirements of a contractor. But the results specification also has a use when no method exists for achieving a particular result, for example in research and development work, or in new (and perhaps hostile) environments. In such cases, the development of a method may be the 'result' of a particular contract, the method itself being available for the client to use in a future contract.

In general, results specifications are simple, and whole processes may be carried out to a specification which states the results in the barest terms, for example:

> The ginning plant shall be supplied and installed to the Contractor's design and shall perform to the following minimum requirements –
>
> (1) throughput 8.33 tonnes of seed per hour;
> (2) capacity of gins 18 bales (as specified) per hour;
> (3) capacity of seed cotton pneumatic conveyance, separators, cleaning equipment and gin feeders 10 bales (as specified) per hour;
> (4) bale press 20 bales (as specified) per hour;
> (5) seed pneumatic system 7 tonnes of seed per hour;
> (6) trash cyclones, conveyor and incinerator 500 kg per hour.

Because of its simplicity, the results specification is likely to place the burden for its achievement squarely on the contractor, and therefore may sometimes become synonymous with lump sum payments. However, the client may well choose a hybrid version of the results specification in which the specifier will state the desired result or results and will also describe essential components of the permanent works. This is very much like the car buyer who wants a car that will carry five people and consume less than 15 litres per 100 miles, but which must have alloy wheels and various other essential features. This concept is an oversimplification of the use and effects of a results specification which, although it may often exist in conjunction with particular payment forms, does not necessarily imply any particular one.

The results specification will leave the pricing of the work very much in the tenderers' hands. Every tenderer will devise their own way of achieving the results and incorporate this in the price. A great deal of what appears above in the discussion of the method of working applies equally to the results specification, if the contractor supplies a method of working. However, in general, the specifier is asking not for method but only that results be achieved. During the tender process the client can build up a more detailed picture of the equipment that will be supplied. Blank data sheets are included in the tender document, to be completed by tenderers, for this purpose. Thus, the tendering process can only change the specification if the contractor actually wishes to modify the result. Modification is valid if the tenderer sees the results as impossible to achieve with his or her own techniques, or as uneconomic to price, or as impossible on the strength of the client's own data. The tenderers obviously have less scope for amending the results than they would have for amending the method.

A specifier who needs to change the result because of the tenderers' replies is in an awkward position. If the result is changed to suit a particular tenderer, then the tenders are not comparable. In such a case it would be best to obtain amended tenders on the basis of the new specification. Unlike the method specification, the results specification does not lend itself to modification in the tender process. It is a more rigid form of specification against which to offer a tender.

If the contractors are expert in the work tendered for, the results specification can produce ingenious and cost-effective tenders. Contractors, after all, exist by making their services more economic than their competitors', and the results specification is the perfect opportunity to do so. Clients stand to gain from a results specification but only if they do not interfere during the contractor's performance. That is not to say that clients must not exercise proper control over how they pay out money, but specifiers should check that their efforts are not hampered by such onerous conditions as will negate any specification benefits. For example, the following is likely to prevent any contractor from producing an efficient performance:

The Contractor shall, before any operation is commenced, submit to the Client his intended plan of operation for approval; which approval may be withheld if the Client decides that the operation will not produce the specified result

If clients get 'cold feet' after producing a results specification, it is clear that either they distrust the contractor or they distrust their own site supervision. It would have been appropriate to have stated:

> The Contractor shall, before any operation is commenced, submit to the Client his intended plan of operation for approval; which approval shall not be unreasonably withheld. The grant of any approval by the Client shall not relieve the Contractor of any of his responsibilities under the Contract.

This would allow the site management to be properly informed without placing too much restriction on the contractor. There are many other ways of producing the same level of control but they need careful thought.

In writing the results specification, the question often arises: what is the result? To specify a cement factory by requiring that its output is so many bags per hour at a given quality is the ultimate results specification. However, it is really only the sort of result that would be written for a government minister who was being asked to approve a project.

In all results there are intermediate stages. Each intermediate stage can have its own results specification. The specifier may indeed specify every bolt in the cement factory; each bolt is a result. But the specifier should not specify the order in which the contractor is to tighten them, for that would be a method specification. In order to maintain a logical progression of thought, the specifier must have already specified (at least mentally) the end-result before setting out the intermediate results. For example, in specifying a bitumen-surfaced road, the specifier should know that the top is to be bitumen before specifying the subgrade; and in a steel structure the specification of the bulk steel should be known before specifying the welds. This is not to say that there are never input constraints – to specify totally from the 'bottom up' often leads to overspecification. By specifying every fine detail of the design as a result, the specifier can also restrict the possible methods of working to such an extent that he or she creates a method specification. The true results specification is one that only specifies the essential needs of the client.

Type of specification

The definition of type, for the purposes of drafting specifications, centres on five ways of presenting instruction. These specification types are:

1   closed;
2   open;

3 restricted;
4 exclusive;
5 negative.

Each type of specification has its own characteristics. It is not necessary that the specifier should present a consistent type throughout a long specification – an impossible task because so many items will be specified. However, in specifying any particular item or group of items, the problem of consistency does arise. In fact, it is possible to argue the consistency case for any part of the specification – from the smallest washer to a whole factory. In reality, specifications divide themselves into areas, or spheres of influence, often by discipline (e.g. instrumentation and electrical, mechanical, heating and ventilation), or into areas in which the client has expertise and those in which he or she is ignorant.

The specifier must first decide on the rough areas of division and then how to apply to each area the five specification types. Discussion and examples appear after description of all the types, because it is not possible to show each individual type without cross-referencing other types.

Closed specifications

Closed specifications are those which, even out of context, offer a complete description of the item. The description will offer no alternative, or any mechanism to apply one. The closed specification is absolute. This absolute character is the key to its nature, and it can manifest itself in numerical or descriptive precision or in the naming of manufacturers or suppliers. When specifying a method (a form of style), the closed specification takes the form of a detailed methodology. Failure to meet its exact requirements provides a major problem for the contractor, and perhaps the client.

Closed specifications can cross-reference other criteria, for example a British Standard. However, to retain their closed nature the standard to which the specifier has cross-referred must also be closed. It is a common mistake for specifiers to think they are drafting closed specifications and then use a cross-reference to a standard which is of a completely different type. Such drafting leads to misunderstanding between the contractor, to whom some leeway has been afforded, and the client, who thinks that the contractor has no choice.

Closed specifications may also refer to a particular manufacturer as being the approved supplier, with no alternative being allowed. This particular approach means that the specifier is in fact nominating a supplier (or subcontractor), and this brings contractual problems in its wake. However, if the specifier is intent on specifying in this way, the source of the specification must at least be current. It is too easy to write a closed specification in such a way that only one manufacturer can comply, only to find that the manufacturer no longer makes the article.

The closed specification is of most use in matching a specification to an existing plant or building, in which case the need for exact duplication is important, or in cases in which there is a need to maintain a low spares range. Local availability may also be a factor leading to the need for a closed specification, but here again, if the specifier does not do proper research of suppliers, he or she may face a problem.

Closed specification can inhibit competition but, on the other hand, a closed specification does cut out fringe suppliers who cannot achieve consistent quality. The specifier must be aware of the effect on price in both cases as, in the latter, the cost of abortive work might be higher than specifying a particular supplier. A valid use of the closed specification also arises when the client requires the contractor to use the client's own products in the course of the contract – perhaps cement or bricks or steel beams, or even soap.

Open specifications

The open specification should not be confused with loose drafting. The specifier must not draft something vaguely, then pretend that he or she was drafting an open specification. To specify the result alone is in fact the same as writing an open specification. However, the concept of the open specification should be quite divorced from that of the results specification, the latter is a style of drafting.

The specifier's use of an absolute set of standards plus the use of samples is a sort of open specification. The specifier tells the contractor the standard to be met, but may allow many source samples to be submitted for approval as meeting the standard. Each sample then becomes the specification. Such a course of action is common in earthmoving work, such as roads or reservoir embankments, in which the specifier's setting of absolute standards is useless if there is no actual source of fill material that complies.

In an open specification the specifier does not name a single supplier of goods, or even supply a list. But the specifier must not specify an item in such a way that only one supplier can meet the criteria (see manufacturers' specifications, p. 59) unless prepared to name the manufacturer. To hide a closed specification in an open one is dishonest – though this may be difficult to avoid where political considerations exist which have made the writing of a closed specification with the supplier's name impossible.

Restricted, exclusive and negative specifications

The two main divisions of type – closed and open – have a number of variations. These are valid subdivisions of type since they do exist in practice and need classification.

The *restricted specification* specifies a range or type within which the contractor can choose to meet the specification requirements. The specifier's use of a list of approved suppliers is an example of restricting the source of materials. Similarly, the limitation of types of material indicates the restricted specification. If the specifier restricts numerical values to a range, or specifies them in tabular form, then he or she is writing a restricted specification.

The *exclusive specification* is of the type in which classes of goods or methods of working are prohibited. The specifier may indeed be prohibiting certain suppliers by excluding a material (e.g. mild steel) or a type of installation (e.g. petrol-fuelled engines), but in most cases a wider choice is left to the contractor than would have been if the specifier had written a closed specification.

The *negative specification* is not strictly a valid type. To write in the negative is very easy, but after a while it becomes tedious to read and irksome to accept. The contractor may rightly wonder whether there is anything that the client will accept. Of course, it may imply acceptance of those items that it does not specifically prohibit, and that may not have been the specifier's intention.

Examples of type

It is not instructive to set out examples in isolation because they are not strictly mutually exclusive. By showing several examples, each type appears in juxtaposition with its relations and it is possible to show changes from one type to another.

Example 1: a closed specification

The pipeline shall be sealed at its open ends, filled with water, and subjected to a 10 m head of water over 24 hours. Leakage from the pipeline over this period shall be nil.

Another example follows, in this case naming a supplier:

The protection shall be applied using Flintkote bituminous coatings of the grades and in the manner specified below. Alternatively, Bituproof coatings manufactured by Shell Composites may be used, substituting the grades exactly.

Example 2: an open specification

The pipeline shall be sealed at its open ends and subjected over a period of 24 hours to a pressure of 10 m head of water. Leakage from the pipeline over this period shall be nil.

A further example follows, showing the omission of the manufacturer's name:

The protection shall be applied using bituminous coatings of the grades and in the manner specified below. The source and type of coating shall be subject to the approval of the client.

The difference between the two types is that in Example 1 the contractor had no option but to fill the pipeline with water and apply the test, whereas in Example 2 the contractor could opt to use an alternative solution, for example to fill the pipeline with smoke under pressure. In the accompanying examples on the bituminous coatings, there is more to the disparity than the omission of the manufacturer's name. The open character of the second set of examples derives from the availability of options.

Example 3: the restricted specification

All materials and equipment supplied under this contract shall be of EU origin.

And another example, on a more technical note:

The Contractor shall only use meters capable of measuring the maximum continuous flow of 13 m^3 per hour with a head loss across the meter of not more than 1 m accurate to $\pm 2\%$.

Example 4: the exclusive specification

If the content of alkali in cement is greater than 0.6% calculated as $Na_2O \pm 0.658 K_2O$, tests using aggregates that meet the specification shall be carried out to ASTM C227. Only cements producing satisfactory test results will be approved.

Example 5: the negative specification

The Contractor shall apply a head of not less than 10 m of water for not more than 24 hours, during which time the pipeline shall not show any leakage. The pipeline shall not be tested in the dry state.

Example 5 achieves no more than the closed specification in Example 1. Being written in the negative, it is somewhat tortuous and pointless – if the contractor is to apply 'not less than 10 m', why should he or she make more work by applying a greater head and risk damaging the pipe? It would be a mistake to think that the specifier had left the option open to the client to demand a higher pressure or a longer test. The words bear no such interpretation. Alternatively, a valid negative is: 'Air entrainment of concrete shall not be used.' This presents a concise prohibition with no hidden implications or subtleties of meaning.

Sources generally

Specifications do not exist without a reason; they are too involved to be written as a speculative venture. So specifiers must appreciate that those specifications that they do not write themselves do have a source – the original writer. The original writer will have been under some constraints, and it is useful for the specifier to appreciate the constraints that applied at the source. Therefore, the specifier should identify the source of any specification that is encountered.

Identification of the source is not enough in itself, the specifier must be in a position to verify the source and depend on its integrity. It is no use stating that a particular source was used if no explanation can be given why that source was used. Sources of specifications divide naturally into five groups:

1   standard;
2   manufacturers';
3   imposed;
4   inferred;
5   one-off.

It is easy to see, in the paragraphs that follow, that these five groups are really influences that are unavoidable in drafting. There is no question of avoiding any of them; the problem is in choosing which of them will be the major influence over the whole document.

Standard specifications

These usually exist in two forms – the in-house and the independent standard. Accepting, for the purpose of illustration, that there are such things as independent standards, they will exist in various forms and derive from various sources – national, international, trade associations, industry, government agencies and the military.

The in-house standard is the easiest to deal with because the specifier will have had some part in drafting it – or will find that it cannot be changed. At least, in-house standards have a clear policy for use attached to them. However, specifiers should constantly make enquiries on the validity of any specification, and be prepared to demonstrate that it should not be used. On the other hand, if the specifier challenges an in-house document, a spirited response is sure to be attracted from some quarter.

Independent standards are more difficult to approach. Such standards are not difficult to apply in the country in which they originate, but they are more troublesome across national boundaries. Specifiers cannot assume that

English-speaking countries will automatically accept standards from the UK. The use of standards across political boundaries is a matter of technical influence. Good examples are the wide use of the US API (American Petroleum Institute) standards in the oil industry, in fact worldwide for welded steel pipelines, the ASTM (American Society for Testing and Materials) standards for gravel roads and British Standards for pipes and fittings. The German DIN (Deutsches Institut für Normung) standards and the French NFA standards (the standards of the Association Française de Normalisation – AFNOR), also appear frequently. However, anyone who has worked in the Middle East and seen the multiplicity of electrical plug-and-socket configurations in a small selection of buildings will appreciate that competition among national standards is both fierce and inappropriate. Specifiers are not always free to choose their preferred national standard because this can be a restriction to free trade for suppliers of engineering equipment and materials. For example, public utilities in Europe cannot prefer one national standard above another within the boundaries of the EU (see p. 35). The various national standards are being given consistency under the control of the European Committee for Standardization (CEN) and the International Standards Organization (ISO). Those British Standards which are now aligned with other European and International standards are designated BS EN or BS ISO.

The specifier should enquire for what purpose the standard was written, and whether the proposed use of it is consistent with that purpose. In the use of standard specifications, too, the question of current validity arises. A quick glance through the catalogue of British Standards will illustrate the fact that standards take a long time to evolve, and some exist beyond the time they should have been laid to rest. Unfortunately, such catalogues do not cite any body of opinion that may have exposed a particular standard as inappropriate, outdated or even wrong. The best the specifier can hope for is a new draft, which might exist alongside the old version.

Even more relevant than the question of whether a standard, in-house or independent, is current, is whether items referred to are still manufactured in accordance with that standard. If there is no current manufacture, then any specification containing the reference is going to be very costly to meet. The in-house standard is prone to the consideration of relevance. If the company is a market leader, then the problem of cost is ameliorated, but if the standard has evolved through 'preferential engineering', without the backing of commercial influence, then the cost of compliance will be high.

Manufacturers' specifications

Manufacturers have goods to sell and spend a lot of money promoting themselves, their goods and their expertise. As in other sources of specification, the quality of the source is very variable, ranging from the advertising copy to serious specifications. The 'copy' end of the range is useless to a specifier as a source, although it may attract his or her attention to goods of high quality. At the serious end of the range there is much good material, and without it everyone would be the poorer.

The manufacturer's specification is bound to be of the closed type – after all, it would be pointless to specify goods that everyone could match. As such, it is useful if the specifier desires to write a closed specification. It is not sensible to turn a manufacturer's closed specification into an open specification simply by deleting a name, especially if tendering for the article is to take place. Tenderers always recognize their competitors' specifications.

The question of the expertise of manufacturers in fitting or installing their items is something of a problem. Although it is possible, and quite usual, for specifiers to write that manufacturers should supply installation mechanics, the quality of that service is usually more variable than the mechanics themselves. This is not entirely surprising as the mechanics usually work singly in a supervisory role over the contractor's workforce. The personal attributes of a single person are all-important, especially if the real expertise is in his or her head. This is particularly true of items supplied to a site abroad, where a change of personnel is very difficult to achieve, and the contractor may have to suffer a manufacturer's ineptitude with no real possibility of remedying the situation. All the specifier can do is be sure to specify that full and complete installation details are supplied ahead of the item, so that, at worst, it can be installed without the manufacturer's assistance. The specifier should ensure that default by the manufacturer in supplying installation assistance does not invalidate the guarantee.

A specifier who intends to use a manufacturer's specification should enquire whether independent test certificates are available on a regular basis. Many manufacturers adopt such a policy, though the independence of the testers may be questioned if the manufacturer owns the testing laboratory issuing the certificates. The specifier should not accept as any more than salesman's spin a once-only test certificate issued for the item range some time in the past.

Of course, the role of the client needs to be taken into account in the matter of the manufacturers' specifications. The specifier should check whether the manufacturers are acceptable to the client.

Imposed specifications

Many areas of our regulated society are covered by rules of one sort or another. No specifier can work for long without coming up against some form of imposed specification other than those 'imposed' by in-house rules (see Standard specifications, p. 59).

Imposed specifications can be direct or indirect, emanate from the client or the law, or be a complex mixture of all sources. The specifier is in a difficult situation because the list of impositions only seems to grow, and ways of meeting them become more complex.

Direct impositions are the simplest source, though not necessarily the simplest to interpret. The UK Building Regulations are probably the most frequently met in everyday life. Any householder planning an alteration will come up against them. There are a plethora of such regulations in each industry, and it is impossible to list them in any single work.

Indirect impositions are those which have general criteria but do not give precise instructions how they are to be met. Items in this category are all those for which a certificate of some sort is required from an independent body to signify that sound design or construction methods (or both) have been used. The Statutory Instrument (SI)[1] concerning oil and gas platforms requires that a verification scheme is established for safety-critical elements before the platform can be operated in a production capacity. However, the criteria for such an issue are under the control of various independent and government authorized inspection agencies. The rules are not found in SI 913, but are those which the industry and the agencies develop over time. Therefore, the specifier is faced with satisfying a set of rules which is a moving target because it is intended that such rules keep pace with developments in a changing industry.

Imposed specifications may also come from purely commercial arrangements. For example, no underwriter will accept a sea transit risk for the transportation of an offshore platform, or a large part of it, until an independent surveyor has issued a certificate that the sea fastenings have been correctly installed and are adequate. The client may also impose his or her specifications, by virtue of the fact that he or she is an interested party (e.g. the manufacturer of the item), or has some political aim (e.g. to encourage local industry), or simply holds a strong opinion. The specifier must know the extent of the client's interest. Some clients actually provide their own specifications for incorporation (e.g. the Turkish Government Water Department (DSI) or the Kenya Highway Authority).

Direct statutory impositions are fairly rare because of the general nature of most statutes, so the specifier rarely needs access to the original law. If the specifier suspects that such access is necessary, legal advice is essential.

Having accepted the existence of imposed specifications, the specifier must check compatibility with those other parts under his or her control. Failure to adhere to the specification requirements can have severe consequences for everyone.

Inferred specifications

The inferred (by the specifier) specification is the hardest to identify. 'Inferred' is used to mean that compliance with/use of is implied from the start of the writing. The inference itself can arise from two sources, namely inferred as an integral part of specifying some other item, or inferred from external sources.

Manufacturers imply the use of their products in some way or other, either directly or through other companies. For example, the manufacturer of a pump may specify the drive train produced by another company, manufacturers of paint may supply complete systems instead of a decorative top coat, or computer manufacturers may recommend software not produced by themselves. The list is endless.

The external influences also derive from many sources: government, client and programme constraints, and compatibility. Governments usually have policies of encouraging local industries; specifiers therefore need to be aware of those industries. Similarly, clients may make subtle suggestions. The actual programme of design or construction may limit what the specifier can rely on because of the state of the market in those items, or because, if methods are specified, it is only those which will allow the job to proceed on programme. A specifier of plant or vehicles must be aware of the servicing and spares networks available to maintain the items. The availability of servicing and spares is extremely important – especially abroad, where the vast range of manufacturers' names bears no resemblance to the very limited servicing available. Specifiers of British goods abroad should not pay overmuch attention to maintaining British content if it means that the job will collapse in a short time through the lack of a proper dealership network.

Money is, of course, an important influence on specification but it is a blanket influence over a whole job. It does not necessarily attach to any particular item or method.

One-off specification

The one-off specification is clearly the ultimate in suiting a specification to its purpose. The specifier who wants to write the one-off may need to determine whether there is another available source. If there is not, then the specifier needs to follow the guidelines outlined in this book.

Effects of size

The size of works is an indeterminate factor; there is no measuring unit which is applicable over the vast range of engineering projects. A 50,000 ha irrigation scheme does not compare with a £50 million natural gas terminal, nor does a thirty-storey building compare with a 3,000 m runway, although they are all 'large'. But the development of a robot manipulator arm is 'small', though no doubt crucial to whatever it controls. Often, in today's terms, size is money, but this can be grossly misleading. Twenty miles of steel pipeline on land in the UK probably costs a quarter of the same pipeline in the Middle East, the difference being made up of factors which have nothing to do with the specification but everything to do with remoteness – shipping being a non-technical factor.

The specifier must approach the task with questions which penetrate the self-congratulatory spin which surrounds projects and ask: what is the client expecting – function or decorative work? How standard is the technology? How much repetition exists? What economies of scale can the specifier achieve? How adequate are the local standards of workmanship? It is these sorts of questions which form the basis on which to structure an approach.

Large works

Large works can be either the sum of many small works or the integral, that is they are either repetitive or not. Definitions (as in the case of small works) are bound to be elusive, but those projects whose commissioning takes place as a single operation are undoubtedly large works if they have absorbed a large amount of resources in their construction. However, this may exclude the integral form of definition in which large works may consist of many single units, for example a large housing project for which individual commissioning is possible and even essential. But all large works share the need for sophisticated management to meet programme and specification. This implies major contractors with large resources, complex testing programmes for materials and methods, and a good deal of on-site testing. Specifiers can rely on there being facilities to maintain the quality level they set. The danger is not so much in overspecifying, but in creating a specification bureaucracy: it can be tested and therefore it must. A multi-million-pound factory building contained the specification:

> HSFG bolts shall be used in accordance with BS 4606 and as follows:
>
> (1) plain and tapered washers placed under part being turned and non-rotating part as appropriate.

Such unnecessary duplication unfortunately appears to occur more frequently on large projects.

Small works

Having decided what makes a large project, the opposite must make a small one. But there are other factors at work: how does one measure intricacy of design or manufacture? Small works, like beauty, exist in the eye of the beholder. Assumptions about the skills and capabilities of small works contractors abound, overseas and at home, and there is a view that the specifier should spell out everything. Such a view destroys any chance of achieving economies of scale in small works. But recognizing greater specialization leads to a lesser need for detail. A hedge layer only needs to be told that the hedge has to be laid to 'Midland style', so it serves little purpose to specify the height of the stakes and the thicknesses of the binders when skills have to be applied to the available materials. Similarly, a contract bricklayer does not need a method specification on bricklaying, but only needs the material and the bond to be specified. What is important is that

Table 3.1 The choices of style, type and source of specification

Style	Type	Source
Method or results	Closed	Standard
	Open	Manufacturers'
	Restricted	Imposed
	Exclusive	Inferred
	(Negative)	One-off

the specialized skills can be demonstrated through some reference work or trial.

Small works pose problems for the specifier. The probable lack of immediate testing facilities is one. Another is the identification in each case of the level at which to pitch inspection requirements. It is with small works that manufacturers' and standard specifications play their largest part in efficient writing. Although there is always a danger of overspecification, the opposite is possible – vague specification. Good libraries of standard specifications should help the specifier here.

Summary: style, type and source

In summary, the choices of style, type and source always exist whatever the size of the work in hand. In tabular form these are shown in Table 3.1.

In large works there will usually be the full choice of styles, types and sources. However, small works will usually rule out the method specification as uneconomic for the client because greater client involvement is required to produce it. Of the sources freely available in small works (assuming that inferred and imposed are not), the one-off specification suffers from the same constraints as the method specification, whereas the manufacturers' specifications probably have their greatest influence. Standard specifications, too, will be very suitable. There is little restriction on the type of specification, because that is independent of the size of works and more dependent on the client.

4   The effect of specifications

'Then you should say what you mean,' the March Hare went on.
'I do,' Alice hastily replied; 'at least – at least I mean what I say – that's the same thing, you know.'
'Not the same thing a bit!' said the Hatter.

The careful specifier

The specifier, as part of the design team and as an employee of the company that may have contracted with the client to provide a service, owes the client a duty to act as a competent member of his or her profession. The notion of competence is not limited to the avoidance of mistakes – even competent people are allowed to make mistakes. But the duty implies that the specifier has a competence in all areas of the discipline in which he or she is offering services to the client. But what of related areas outside the expertise of the specifier? These are very varied, but usually include insurance, economics, law and the definition of legal liabilities, as well as other areas of technology which may themselves be highly specialized. It is necessary for specifiers to be aware of their limitations and to seek specialist advice for those aspects of the work outside their competence. The specifier's employer still carries the responsibility for the advice obtained from others, but is usually able to fall back on a similar duty owed to him or her by those advisers. Of course, if the specifier is an employee of the client, then the specifier owes no further duty than that contained in the contract of service with the employer, namely to act in a responsible fashion as an employee.

Although the specifier may owe a duty of care to the client during and after the writing of the specification, no such obligation is due to the eventual contractor or any of the tenderers. The specifier has no implied relationship (in the legal sense) until becoming part of the site team, and then only as a result of the contractual relationship between the client and the contractor if that contract makes provision for the client (through the site team) to act fairly and reasonably. The absence of a duty at the specification stage does not mean that the specifier is in a position to mislead tenderers by deceitful drafting. In fact, by doing so, specifiers would fail in their duty to the client or to their employers. Though in most offices the element of deceit is thankfully absent, it is sometimes the case that specifiers must appear to be more certain than they may feel is prudent. For example, to specify a result that the specifier believes is unachievable in the conditions at the site is misleading because the tenderer will put in a price on the basis that the result is achievable; the result will be a claim. At the same time, the specification is no place for gratuitous advice to the contractor who must be assumed to be competent in his or her field.

The influence of standard on cost

The specifier, in common with the designer, should be constantly trying to attain 'appropriateness' in the level of specification; there is no such state as 'money no object' in the specifier's mind. If specifiers are under that constraint, then they must specify to the highest quality levels, itself an important restriction. The relationship of cost to quality and time can be shown in a simple diagram such as Figure 4.1.

One problem is that costs appear differently to different clients. A client who can afford a high capital cost for, say, a pump may well be hoping to recoup that cost through higher efficiency and lower running costs. Another client may be able to raise money for maintenance easily and therefore may accept higher running costs if the capital costs can be kept down. Typically, the capital cost of a pump might represent only 15%, or less, of the total cost of owning and operating the pump for the whole of its life. The design life of an item of plant affects its depreciation rate, thus the design life becomes the concern of the specifier. The compatibility of design life is important, i.e. matching components in a plant, so that the shorter design life items do not prejudice the safety or the operation of the plant as a whole. In general, clients look at the total cost of a project over its whole life. 'Whole-life costing' often forms part of the tender assessment and the information sought

Figure 4.1 The conflicting influences of time, cost and quality on specifications

from tenderers to meet the specification will be used to calculate the total cost of the proposal over the expected project life.

Costs are often incurred quite separately from any question of quality. One system design may call for a higher import content than another design of otherwise equal quality; in countries with exchange control limitations this presents a cost problem. The use of common materials usually keeps down the price, as does specifying an off-the-shelf item. But to specify the common item and then call for some minor modification may destroy the initial price advantage. In another setting in which the specifier may call for an item that needs specialist installation advice or supervision this may not be directly reflected in the item cost but may appear elsewhere as a hidden cost, perhaps in the general overheads on the work. The specifier should check that the hidden costs are properly allocated to items affected.

The great danger is of 'preferential engineering'. Adding to a basic specification to the extent that money is no longer being used efficiently is a constant threat, and the specifier should always work to a basis of achieving the performance or other goal and no more. In the end, it is the client who will pay, not the specifier.

Approvals

There is a temptation to insert in a specification a great number of 'approvals' stages. This is especially true of those writing a small part of a large specification for which they do not know, or are inclined to doubt, the provisions that the collator of the various specifications has made for on-site approval.

In many contracts there will be general clauses which are designed to prevent work being covered up or hidden before it is checked. Provision for approvals is not just a problem of contract drafting; it is one of management. The specifier should assemble a list of the formal approval stages which have been specified in order to see whether any rationalization is possible. A process of checking, and possibly double checking, does not fit easily in a contract based on partnering principles, in which the parties have agreed to work in a spirit of trust. The approval stages will occur, but through meetings and consensus. It is vitally important that the approval stages are appropriate to the relationship between the parties and the needs of the contract.

The existence of approval checks is very important when specifying for a design contract. A balance is required to be struck between interfering in the legitimate design process and the need to ensure that effort is not being wasted. To specify that 'doors shall be capable of opening without striking fixed furnishings' is quite unnecessary as the provision of adequate doorways is implicit, but to specify that 'doorways shall be not less than 1.75 m wide' would be legitimate even if no reason were given.

Problems arise when the validity of approvals is questioned. Approvals which vary the contract have no authority unless subsequently ratified by the contracting parties. Approvals which are incorrect in terms of the contract rely for their validity on the exact words in the specification. The specifier must be explicit in the wording of approval clauses as to their exact scope and the degree to which they bind the client and the contractor. For example, if concrete is specified to a minimum cube strength of 30 N mm^{-2}, then samples which exhibit less than 30 N mm^{-2} imply that the contractor is apparently in breach of contract, even if no formal condemnation comes from the client. However, if concrete is specified to have a cube strength of '30 N mm^{-2} or other suitable value as decided by the Engineer', then a value of 28 N mm^{-2} or even 10 N mm^{-2} would satisfy the terms of the contract if the Engineer gave approval.

Approvals wrongly given and technically incorrect open up a huge field of argument that dismays lawyers. It is necessary therefore to ensure that clauses requiring formal approval have the method, scope and technical standards clearly defined. In the end, the only binding approval is the final certificate subject to the conditions of any existing warranties.

Approval of the engineer

In contracts in which there is a client's representative who has specific powers (often the Engineer) the phrase 'approval of the Engineer' occurs all too frequently. The reason is usually some uncertainty on the part of the specifier; but there can be valid reasons for such a phrase, such as the Engineer giving technical approval by means of comparing the work with the specification. There should be no need for the phrase 'all to the approval of the Engineer', unrelated to any specified result. However, approval may follow a specified result, e.g. where the use of the phrase provides the means for establishing exactly when the reinstatement is complete (to the 'original condition'):

> Tracks may be used as haul roads by the Contractor provided that such tracks are reinstated to not less than their original condition to the approval of the Engineer.

However, note the difference in the following example:

> The Contractor shall supply roof tiles to BS EN 1304 or to the approval of the Engineer.

The specification provides a choice, and it will be satisfied if the tiles conform to BS EN 1304 or if the Engineer approves them. This allows the Engineer to approve tiles which may be far inferior to the specified tiles, and he or she can do so without varying the contract between the contractor and the client, because by exercising discretion the Engineer satisfies the contract. The question to be asked is whether the specifier intended such wide discretion, especially if the Engineer has delegated such powers to quite junior site staff. The uncertainty of the quality of supplies may have been good reason for such a way of specifying the tiles, but a better way of phrasing it, which, at least, provides some sort of minimum standard for the site staff while allowing flexibility in interpretation, would have been:

> The Contractor shall supply roof tiles to BS EN 1304 or tiles of equivalent quality to the approval of the Engineer.

The example above illustrates again the inadvertent delegation of specification to others by the inappropriate use of words. There can be a more important effect, namely uncertainty at the tender stage. In the first example the tenderer is presented with a problem: is he or she to price the tiles at, say, a high imported price in order to meet BS EN 1304, or is the tenderer to gamble on the Engineer approving local tiles which do not conform to the standard? Such uncertainty should not be considered as 'contractor's risk' and, in any case, it is doubtful whether the contractor is obliged to provide a more expensive alternative without additional payment. If the specifier is really uncertain of the relative merits of the two sources of tiles and the contract form allows for flexibility in pricing, then, for example, two alternative items should be allowed in the bill of quantities:

> 1.11 Supply and fix roof tiles (local source)
> 1.12 Extra over to item 1.11 roof tiles to BS EN 1304.

Thereby some certainty is introduced into the price.

Responsibility to the client

As well as affecting the tender price unwittingly, the specifier may affect the client's costs in claims. In a purely construction contract a clause such as the following might refer to the design of the earthworks, for which the contractor had no responsibility:

> In the event of slips and falls occurring the Contractor shall make good all earthworks and associated damage and carry out any requisite modifications of the Works to the satisfaction of the Engineer, without any additional payment.

If this is intended to cover more than poor workmanship, it will fail. The specifier has to remember that the contractor can be held responsible only for that which he or she contracts to do. Any attempt to add sweeping statements such as 'without additional payment' will fail – unless there is a contractual obligation to do whatever is being qualified in that way, in which case the qualification is unnecessary anyway. Also, in some standard conditions the above liability would be specifically excluded if it were due to a design fault or unforeseeable ground conditions. To attempt to protect a client in this heavy-handed way means failure in the specifier's duty to give the client proper service and advice.

Responsibility to the contractor

In parallel with the specifier's direct responsibility to the client is the responsibility to the contractor to produce a document which is clear and unambiguous. To do anything else affects both the client and the contractor. A clause such as the following implies (or even states explicitly) that the contractor has to allow in the contract price for all changes arising from 'further drawings and orders' from the Engineer:

> The Works shall include... to the true intent and meaning of the Drawings, Specifications and further drawings and orders which may from time to time be issued by the Engineer.

Presumably, what the specifier means is that the contractor has to carry out the orders of the Engineer in so far as they explain or clarify the drawings to which the contractor has tendered. In any case, this clause is a dangerous duplication (for all duplication is potentially dangerous) if the conditions of contract contain a clause like the first part of the FIDIC clause 3.3 (FIDIC Conditions of Contract for Construction):

> The Engineer may issue to the Contractor (at any time) instructions and additional or modified drawings which may be necessary for the execution of the Works and the remedying of any defects, all in accordance with the Contract. The Contractor shall only take instructions from the Engineer, or any assistant to whom the appropriate authority has been delegated under this Clause. If an instruction constitutes a Variation, Clause 13 [*Variations and Adjustments*] shall apply.

A similar example of a clause, like the 'further drawings' example above, which binds the contractor but which is unfair, and obviously arises from indecision in the mind of the specifier, is:

> The location and the number of canals and drains and the number, location and details of the structures shown on the Drawings are subject to amendment by the Engineer and will be confirmed by the Engineer during the course of construction.

Such a clause, considering that a tender will have been invited on the basis of the drawings, offers the client carte blanche (probably leading to a justifiable claim) to make changes at will. If the client requires this degree of flexibility then the form of payment should be chosen accordingly (see Figure 1.1, p. 22).

In the end, because of the lack of a formal legal relationship between the specifier (as opposed to the client) and the contractor, the specifier's duty to the contractor has to be one of fairness, both as a moral obligation and in terms of self-interest in order to obtain a price for the work that will stick.

Workmanship

Specifiers tend to work at extremes. They assume either that the contractor is virtually clairvoyant or is an idiot who needs to be told where to place every nail. This tendency is most apparent when it comes to workmanship: how far ought it to be specified, if at all? It is important to remember that, if the contractor has contracted to produce a result, the workmen used to achieve that result are of the contractor's choice. Probably the limit of the client's influence would stop at the approval (and the possible withdrawal) of the contractor's relevant manager and particular specialists.

Workmanship also exists in its own context. First-class workmanship in one country will often be to a different standard to that found in another. Is workmanship therefore to be 'first class' or 'the best available locally'? A clear distinction between the two in the specification assists the tenderer in pricing for labour and in the subsequent management of the contract. For the specifier, it is no good specifying 'first class' if the prices quoted are unacceptable to the client because the tenderer has had to assume an input from expensive imported personnel.

Much of the standard of workmanship is specified indirectly, e.g. verticality, clean edges, even slopes, smooth transitions, and so on. An example in which an incorrect assumption was made by the specifier is to be found in the following road specification:

> The whole of the earthworks for the several parts of the Works shall be carried out to the dimensions and levels shown on the Drawings, or other such dimensions and levels as may be ordered by the Engineer.

During construction its subsequent alteration is specified as:

> The whole of the earthworks for the several parts of the Works shall be carried out to the dimensions and levels within the tolerances shown on the Drawings, or to such other dimensions and levels within the tolerances as may be ordered by the Engineer.

The above alteration was necessary because the accuracy, being a measure of workmanship, required by the original specification allowed for no error or deviation at all – hardly practical in a road.

Inconsistency

When specifications are drafted by a variety of people, usually specialist departments, inconsistencies are bound to creep in, occurring between the various parts of the specification, e.g. the drawings, conditions of contract and pricing sections. In all cases inconsistency can be overcome by planning with others (see p. 29).

Apart from the problem of drafting inconsistencies, there is the problem of inconsistencies in the items specified. Are the levels of quality consistent? Has the specifier asked for high-quality cabling and low-quality switchgear? Has he or she specified spark-proof electrical fittings in a rural pumphouse? All faults more common than at first may be realized, one can see them every day in buildings.

What is the effect of inconsistency on the price? Any tenderer faced with inconsistent specifications will play safe and opt for the cheaper solution every time, then rely on a future claim to raise the price for that item. Unless the tenderer can resolve these questions during the (usually short) tendering period, there is no real alternative.

Once work has started on the basis of an inconsistency, the site staff will be faced with the problem of trying to decipher the true intentions of the specifier. Often the result will be a compromise, which may not be in the best interests of the client.

Errors

In general, the contractor is protected if the design or specification is wrong. However, as in all areas of liability, there are uncertainties, but errors can be categorized into three classes: the incorrect, the partially correct and the ridiculous.

When a specifier writes '10 mm reinforcing bars to all stairwalls' but means '16 mm bars', or '240 V' rather than '480 V', then the contractor can carry no loss. The contractor is bound by the contract to supply 10 mm bars, and the contract will have to be varied with the contractor being paid extra to supply the correct bars or the correct transformer.

If a specifier states 'three holding nails per tile' but the specified tiles have only two nail holes, then the contractor should bring his or her implied (in the legal sense) expertise to bear by pointing out the supposed discrepancy, but, of course, it may be that the specifier does require a third nail hole to be drilled in each tile. The ridiculous in 'fire extinguishers all to BS EN 60598' (the astute will know that the standard applies to lights) is so plainly wrong that any contractors who install lights instead of fire extinguishers have only themselves to blame.

More serious are the cases of 'error by approval'. In many design contracts the client's staff approve drawings before items of work can commence. What is the position if the approval given is incorrect? Again, this is a complex problem of liability, but, in general, most standard forms of contract protect the client in such a case by wording such as: 'No approval... shall act to relieve the Contractor of any of his liabilities under the Contract.' Here the effect is that, even if a beam of insufficient strength were approved in a drawing submitted by the contractor, the contractor would still have an obligation to provide a beam to take the specified load.

The last form of error is, perhaps, not an error at all. This is where the specifier drafts a specification that conflicts with 'normal trade practice'. It usually happens because the specifier leans towards theory more than does the contractor. Contractors have successfully pleaded 'normal trade practice' as a defence to allegations of failing to meet the specification.

Specifications in subcontracts

When writing the specification for a subcontract, the specifier must ensure its consistency with the main contract. At the same time, subcontract specifications are often more general than the main specification because the details will often be provided by the specialist subcontractor (as intended), and the specifier relies on the subcontractor's design and drawings. Consistency cannot easily be achieved by cross-referencing between the main contract specification and that of the subcontract; indeed there are unresolved legal complications in any such attempt. These complications are discussed in works on contract law, but the specifier can avoid these difficulties by making the subcontract specification a complete entity in itself. Appropriate abstracts can be taken from the main contract for inclusion in the subcontract specification, together with new specification clauses to cover any additional needs. If this course of action is adhered to, the writing of the subcontract specification can follow the same methods as used for the main specification.

As subcontract specifications often rely heavily on the approval of drawings submitted by the subcontractor, this process becomes one of 'specification by delegation', from the main contractor to the subcontractor. The specifier should ensure that the overall specification is not being altered as part of a bargaining process between the main contractor and the subcontractor.

Nominating suppliers and subcontractors

In some cases the specifier will identify particular suppliers or contractors who are the only acceptable source of goods or services and instruct the main contractor to use them. In such cases, the client is said to 'nominate' the supplier or contractor, and the main contractor is then told to enter into the necessary contract. Out of such an arrangement comes a difficulty of compatibility between two contractors, namely the main contractor and the nominated subcontractor or supplier (it makes no difference).

Here the problem of compatibility is not the same as that arising out of the case of two contractors on the same site, both of whom are responsible to the client direct. In the case of nomination, despite it being the result of an instruction from the client, the responsibility of the subcontractor is still to the main contractor. Therefore, the specifier needs to define the boundaries of the specifications for each to avoid disputes between the main contractor and the nominated subcontractor, always a possibility and especially if the main contractor permitted the nomination against his or her 'better judgement' under pressure from the client.

The same principles apply to supply-chain suppliers and subcontractors, for example in the contractor's use of a client's framework supplier for, for example, valves. However, they are less likely to lead to a dispute between the contractor and the supplier in such a relationship because the supplier is fully aware of the client's requirements and will be able to provide valuable advice to the contractor at the tender stage. The framework supplier also has a keen desire not to cause knock-on disputes to the client because of the possible damage to the long-term trading relationship between the supplier and the client.

Other contractors on site

On many large sites there is often more than one contractor (i.e. excluding subcontractors of the main contractor). If this is likely to be the case, then the specifier should recognize it and lay out the specification accordingly.

It is important to show in a particular contract where that contractor's responsibilities end. The rest can be referred to as 'work done by others'. There should be no contractor-to-contractor responsibilities within the specification. If each contract has been drawn up to provide the same clear boundaries, any interference that cuts across those boundaries should give rise only to identifiable claims, e.g. in a case in which a power generator supplies an incorrect voltage to compressor motors installed by the compressor manufacturer. Provided that the specifier has correctly specified the outgoing and incoming voltages in each case, the claim for motor damage can be identified within a single contract.

Obsolescence in specifications

Specifications are in a constant state of growth when viewed over a whole industry. New materials, new regulations and variations in marketing by large, important suppliers or contractors all produce changing specifications (or should do). In writing any specification, the specifier must be aware of recent changes in those parts of the industry involved. Specifications often become fossilized quickly. This is only natural – people like to stick to familiar ideas.

If specifications are written using past versions, the result may be out of date, yet when the specification becomes incorporated in a contract it will be binding on both parties. The fact that a specification is obsolete, assuming that it is not illegal as a result, is not relevant. Clients are perfectly within their rights to follow an out-of-date practice if they think that it is to their best advantage. For example, there may be large stocks of items available to a withdrawn standard and so it may be commercially sensible to specify them. In some cases, standards are not replaced and the manufacturers are left to their own devices. In fact, the obsolete standard may be the only relevant specification.

Obsolescence on the part of others (e.g. manufacturers) is not binding on the contractor, but the result may be more expensive than if using the item specified. The choice is whether a partial redesign is necessary to accommodate the change in manufacturer's specification or whether the manufacturer should be asked to provide 'specials' that conform to the original specification.

Should there be any changes to the specification as a result of obsolescence which occurs, or only becomes apparent, after a contract has been signed, it can only be incorporated in the contract, as any other change, with the consent of both contracting parties. The consent may have been given in advance through the parties' agreement to the contract variation clause. Model specifications (as discussed in Chapter 7) have an in-built obsolescence by virtue of their common use. In fact, any specification which has been in use for a long time should be treated as suspect unless there is evidence of a good updating system to incorporate necessary changes.

The tender

The first chance for the specifier to see the result of his or her drafting is when the tenders arrive from suppliers or contractors. In construction tenders the effects may not be immediately apparent because of the interwoven complexities of the work. But in simpler procurement contracts the deficiencies in wording are brought home forcefully. A specification for agricultural equipment which called for an '8-row ridger with Lister ridging bodies set on a tool bar at 1 m centres' brought a variety of responses:

- an 8 m ridger with 8×0.9 m gaps with nine bodies;
- an 8.4 m ridger with 8×1 m gaps and nine bodies;
- a 7 m ridger with 7×1 m gaps and nine bodies;
- some with single-bodied ridgers at the ends and some with double.

The reason for these variations is that the specifier had failed to make clear whether a results specification was being written for a ridger to produce eight rows in one pass, or a machine specification for a machine with eight ridger bodies on a tool bar. The result was that, while the tenders were being evaluated, a good deal of clarification was necessary in order to ensure that tenders eligible on price and other criteria really would produce the final field configuration.

Great variations in the tender price may be indicative of poor understanding of the specification, not always the fault of the tenderers. In construction tenders the large variations may, of course, be due to simple variations in the evaluation of the risk element. But if single items vary widely in price, the risk element may be found to be inherent in the way the specification was written. Also, it is probably true that if a tender contains a plethora of separate memoranda qualifying the offer with respect to technical matters, then the specification may be at fault.

As for amending the specification during the tendering period, either because of apparent errors or because of enquiries, this should be done with great care. It may be preferable to allow the tenderer to draw his or her own conclusions from the wording than to tinker with the document. The period allowed for tendering is usually very short and does not allow for a thorough revision of both documents and drawings. It is better to learn the lesson for next time, though each case will have to be tackled on its own merits.

Effect on tender price

The contractor will only supply goods to the quality specified, rarely to a better quality and, hopefully, not to a lesser one. Quality is as much part of design as are dimensions and performance. If the designer has properly calculated what is required, then the specifier must be aware of the facts. It is very easy for the specifier, perhaps in a different office from the designer, unwittingly to increase the cost of meeting a design. This is most easily done by introducing some sort of standard into the wording, e.g. by adding the words 'all to BS...'. This will usually mean that the specified item will be more expensive than the designer's original idea because it costs money to comply with independent standards in terms of production checks, tooling, and so on. Many perfectly adequate items are produced to no standards, but to commercial quality; but it does require engineering judgement to decide the most suitable. A retreat into the words 'all to BS...' should not be allowed as a substitute for real judgement because the client will have to pay a higher price than is necessary for the finished product. However, the specifier must always be aware of any applicable regulatory requirements, such as CE marking, that would impose a quality standard on the specification.

Bills of quantities

Bills of quantities, though usually the work of quantity surveyors, are often a sort of specification, and the specifier needs to pay attention to how the bills are laid out. It is quite feasible to have contracts without specifications appearing as a written body of text. The descriptive item in the bill format may fulfil this requirement very adequately; an example is given below:

Item	Description	Qty	Unit
13.01	Section 13: Sanitary fittings White vitreous china WC suites, comprising closet pan to BS 3402 with floor fixing bolts, plastic seat to BS 1254, 9-litre flushing cistern with level handle and ball valve to BS 2456, unplasticized PVC flush pipe to BS 1125: supplied complete ready for installation	20	No.
13.02	Install 13.01 per Drg 37211	20	No.

A briefer example fulfils the same purpose:

Item	Description	Qty	Unit
14.04	Aluminium lever sets to BS 5872	160	No.

Both of the above examples rely heavily on the words abbreviated by reference to a standard. In effect, the descriptions are very detailed. This need not always be the case, as the following example illustrates:

Item	Description	Qty	Unit
18.02	Motorized inlet fan for wall mounting rated at 300 m^3 h^{-1} and complete with filter and control unit	2	No.

In another case the item specified is so common that no description is necessary, or its precise identity can be deduced from previous items in a system:

Item	Description	Qty	Unit
20.41	15 A plugs	300	No.

But, as in all specifications, there should be no repetition, as here:

Item	Description	Qty	Unit
5/103	Supply and fix 500 × 1500 mm vermin screen all as specified in Clause 2.09	43	No.

The words 'all as specified in clause 2.09' are unnecessary because the item, in any case, should have been specified uniquely. The repetition is dangerous because of the cross-reference – it may be that two types of screen are required. If the cross-reference is incorrect, then too many of one item will be bought and none of the other. Cross-referencing from the bills to the specification should be avoided because they may well be done by different people at different times, and the addition of a clause may result in 'clause 2.09' being 'Gate valves 3 in. and under'. It would have been much better to have written, say, 'aluminium vermin screen' in the first place.

Another example highlights the same problem in a different way:

Item	Description	Qty	Unit
110	Supply and lay dry stone pitching 0.20 m thick on level or sloping beds including 0.15 m gravel backing	1,100	m²

In this case, the clause in the specification describes two types of pitching, so that 0.20 m pitching was to be laid on 0.10 m backing and 0.30 m pitching on 0.15 m backing. It would therefore have been better to have classified two types of pitching by calling them 'Type 1' and 'Type 2'. The description would have read: 'Supply and lay dry stone pitching Type 1', thus avoiding imperfect repetition and errors.

Cost estimates

It may not be apparent how cost estimates come into the business of writing specifications. But the act of writing a specification is a major influence on the final cost of the work. Prior to the final cost estimate, there will have probably been other estimates: a preliminary at the feasibility stage, and then a budget estimate submitted to the client. A specifier who is not aware of the bases of any previous estimate may, by applying inappropriate criteria, make a nonsense of them. Even if the estimates, design and specification are all written within the same document, there may be differences in philosophy between the people responsible for the various sections. The situation is exacerbated when many different firms are involved in the design process, as in a consortium. Therefore, the specifier, along with everyone else involved, should be aware of the overall philosophy and objectives on which the cost estimates will have been calculated.

Measurement

In deciding how much a contractor should be paid at any stage in the construction process, many clients rely on quantity surveyors to carry out the measurement. But although some people consider that measurement is the province of the quantity surveyor alone, many clients and firms do not employ them, and so the specifier should understand the way measurement is specified and carried out. Specifying a method of measurement and how it is applied in detail is part of the specifier's job. It may be that all that need be done is to write: 'the Civil Engineering Standard Method of Measurement (CESMM) shall be applied to this work' or 'measurement shall be carried out to the Ministry of X method'. However, if a standard method is not being applied, then all the items to be measured must fall into one of the specified categories.

The way that an item is measured will affect its price in relation to another item, but the overall effect on the contract sum should be one of cancelling out. For example, in building an earth embankment, measurement is usually on the final 'in place' volume, that is no allowance is made for the swelling and the subsequent shrinkage of the soil. This will produce a higher rate per unit volume than if the measurement was based on the volume taken out of the borrow pit. However, the quantities in the latter case would be higher, though at a lower price. This illustrates the price effect of one particular risk – soil shrinkage. Whatever way the work is measured, the contractor will want to be compensated for doing the work involved. The choice of method of measurement should produce certainty as to how all the items are to be priced in the initial tender and how they are to be paid in the contract price so that the finally calculated contract price will be as near as possible to the original tender for the same work.

However, sometimes certainty is an impossible target. In the embankment example the ground conditions may be so variable, or unknown to a potential contractor, that it is impossible for the contractor to make an allowance in the price for the excess of cut over fill, thus the client will have to take the risk of the amount of the excess by measuring only the cut in the borrow pit.

It is tidier if the measurement details are not spread out through the technical specification, but rather put as a preamble to the bill of quantities. This has the advantage of allowing specifications to be used separately from the measurement criteria in cases in which the latter may change from client to client but the former remains the same. The counter-argument, that if the specification changes, then this may affect the method of measurement, and keeping them together ensures consistency, is a valid one, so the specifier should make his or her choice and be consistent throughout any particular contract.

In a measured-rate contract the purpose of measurement is to present to the client the breakdown of how the contractual liability to pay is itemized. It is not a payment method, as such, and the verb 'to pay' and its forms should not appear in the measurement sections. The measurement may be used by the client and the contractor as a basis for the payment application, but the contractual liability to pay does not purely concern physical dimensions of work done; it may involve currency exchange rates, escalation of price, repayment of advances, allowances for materials on site, and other similar items. Clauses covering such items will usually be in other parts of the contract documents.

The method of measurement should be chosen or written before any taking-off of quantities from drawings begins. This is easily done if the designer, specifier and quantity surveyor use a standard with which they are familiar. If a special one is to be prepared, it is important to ensure that at least a draft reaches the designer in the early stages of the design to avoid incorrect billing of items. The designer can add to the draft any items for which no method is specified, and when the specification and drawings are complete, the method of measurement can be checked against the final taking-off and writing of the bill items.

Before attempting to draft a method of measurement, the specifier should study the various standards that are available in the industry and use them where appropriate. If a 'one-off' method has to be written, the standard can be used as a checklist of what to include.

Operation and maintenance manuals

On many projects it is necessary to hand over to the client the information that explains how to operate the project as well as to provide the client with the necessary record drawings, the latter being normal practice. It is normal practice to provide operation and maintenance manuals to the client on any project which includes a process. The client may also have appointed a separate contractor to carry out the operation of the project after commissioning. In this case, the operation and maintenance manuals would be provided to the client for use by the operating contractor. Whether or not to specify operation and maintenance manuals in detail will depend on the merits of each case. As a general guide, the manual should be specified in some detail if any of the following apply:

1 the client has little experience in that field of operation;
2 the project outcome is a complex operating process;
3 there are legal requirements that manuals should exist;
4 the cost of producing the manual is likely to be significant.

In the production of the manual the responsibilities of the various compilers and the contractor may overlap extensively, or may only fall on one or the other. If the employer's Engineer is to produce the manual as part of his or her brief, then the Engineer will obviously obtain a great deal of information from the contractors and suppliers involved. They must be informed at an early stage what is required of them. To assume that the Engineer is to produce the manual and that there is no need to specify its contents is wrong. The site staff cannot be expected to produce it out of a hat at the end of the construction phase! At the very least, the specification should identify the party which is to produce the manual, its general contents, and the actions required of the contractors and suppliers.

If the manual is to be produced by the Engineer, then he or she will need to assemble details of the separate items which constitute the project, and a clause such as the following will suffice:

> The Contractor shall furnish to the Engineer before the Works are completed those operating and maintenance instructions, together with drawings of the Works as completed in sufficient detail to allow the Employer to maintain, dismantle, reassemble and adjust all parts of the Works.

If producing the manual is to be part of the contractor's work, then the specification needs to be presented in greater detail. This is especially necessary if the cost of producing the manual is likely to be significant because the contractor, at the tender stage, will want to price the work involved and include it in the bid. It is entirely wrong to imply in the specification that the manual is unimportant or general in content if, at the end of the construction period, the contractor will be presented with a request for details which may not even have been recorded or are hard to trace. This situation will be exacerbated if subcontractors are involved, who may well have long departed.

Samples

The use of samples is a tempting method of specification, especially if the specifier is in doubt, but it can often be seen from such specifications that the writer is in doubt.

The specifier can rely on samples in two ways: first, to provide an exact model which all future items will replicate; and, second, to provide a guide which all future items will follow as far as possible. In the first case the specifier might state:

> ... sand-coloured stone to walls... colour to be chosen from samples provided by the Contractor.

In this example, when the client has approved the sample, both client and contractor use the sample as the ultimate test of suitability. Similarly, but more precisely, the specification might call for:

> ... holding down bolts to BS 7419, the Contractor shall submit samples for approval.

Here the contractor knows that, having obtained one approved holding down bolt from the factory, it will be possible to obtain any number of identical holding down bolts. In the case of the sand-coloured stone there may be variations in colour in the quarry area; indeed the source of the stone may become exhausted. In such a situation the sample is irrelevant and approval must be sought for a new sample.

The second case is to use a sample to test a source against a precise specification which may itself be impossible to attain. For instance, a sub-base road material may be specified as in Table 4.1.

It may be impossible at the design stage for the specifier to know whether the specification can be met completely because of the limited information it is possible to obtain from identified borrow areas. If the specifier then adds 'Samples to be submitted for approval', this then raises a doubt as to whether the sample (once approved) becomes the specification or not. In addition, there arises the question of what happens if the contractor is forced to change the source of supply. Should the contractor submit another sample against the original sample or be constrained to meeting the original specification? To overcome the possible conflict, the specifier should state clearly that the specification, not the sample, is the starting-point for any new material, for example by saying:

> Approval of a sample from a source does not imply approval of all material from that source, and if any new source is necessary, the Contractor shall supply a sample for approval by analysis and comparison with the specification.

Of course, not all samples are picked off the shelf or dug out of the ground. The 'constructed' sample, for example a wall or a weld, is a useful way of establishing a check on the achievable standard of workmanship. In the case of the welding procedures, or the compaction procedures for plant on earthworks, it is not the specification that is being set by the sample, only the method of its achievement.

Prototypes

The words 'sample' and 'prototype' may be interchangeable in certain situations, so the specifier should be clear in his or her mind which is required.

Table 4.1 Example of a specification for a sub-base road material

BS sieve size	% material passing through sieve
3 in.	100
2 in.	80–100
1 in.	55–90
3/8 in.	35–65
1/16 in.	25–55
No. 7	20–45
No. 36	15–25
No. 200	5–15

A good separation of the two ideas is that a prototype is specially made and would be paid for as a separate item, whereas a sample is not so much 'made' as 'supplied' and so does not need separate payment.

Inspection services

Separate contracts can be let to specialist firms which will inspect and test items before they are dispatched from the manufacturer. In civil engineering, this usually includes plant and equipment specially manufactured or adapted for the particular works. In the case of a pump station the manufacturer of the pumps will probably be a nominated subcontractor or supplier. In addition, the pumps will probably be manufactured many miles from the site and the client will have to rely on test certificates supplied by an inspection agency. Of course, there is nothing to prevent clients from carrying out their own tests, but in most cases this is not done.

The use of the inspection agent poses special difficulties for the specifier, unless he or she knows beforehand how inspection agencies work. To write a detailed method specification for an inspecting agent may inhibit the use of the agent's relevant experience, or may be subordinated to the agent's experience and ignored. In any case, it is rarely possible to 'inspect the inspectors', only to rely on their certificate. The solution is to write a full specification as usual for the contractor and send a copy to the inspecting agent. This leaves the method of inspection to the agent. Specifiers should not be tempted to leave the specification vague in the hope that the inspecting agent will write the details for them.

Inspection is a high cost to the client. It should not be specified unless the value of the inspected item warrants it or there are no manufacturers' certificates available, which can be used as a form of guarantee, as is the case with bulk materials.

Annotated specifications

The certain way to achieve clear specifications is through the unambiguous use of words. It is a sad fact that many documents need 'guidelines for interpretation'. A specification written in the warm certainty of the office may evoke a different response on site, where decisions need to be made quickly. The advent of word-processing software (see Chapter 6) makes selective insertions in text very easy. One idea to aid the designer's site staff to interpret what is required is to have the site office copy of the specification annotated by the specifier. This copy of the specification would not form part of the contract documents, but would allow – for each clause if necessary – notes explaining why the clause was written in a particular way and what the site staff are to understand for its interpretation.

The need for annotation is not in contradiction to the need for clarity, but it allows the specifier to put forward reasons why one particular method or item was chosen instead of another. This approach is not intended to promote more guidelines, but to explain to site staff what the design policy was when the specifier wrote the clause. However, a difficulty with this approach (see Table 7.1, p. 151) is to know when to stop and allow the site staff the freedom of their engineering judgement.

5 Writing specifications

'When I use a word,' Humpty Dumpty said in a rather scornful tone, 'it means just what I choose it to mean – nothing more nor less.'

How to start

The basic questions that face all writers apply to the specifier as much as anyone else: What do I want to say? Who is the reader? How will I say it? There is no escaping these three questions. Someone, somewhere, has to read what the specifier writes. If the specifier does not write what the reader needs to know, the reader will remain uninformed. If the specifier does not write such that the reader feels that he or she is actually being addressed by the specifier, then the reader will not react positively to the information. And if the specifier is unclear as to how to put over the information, the reader will be confused.

What to say is usually the least of the specifier's worries when writing technical documents. Usually specifiers want to say everything at once, which prevents the transmission of the information in a logical fashion. When it comes to the general specification, often describing the workscope, then the problem of what to say becomes more difficult. The specifier should quickly identify the part of the document that he or she is to write and maintain a compartmentalized approach throughout. The earlier chapters in this book show how to deal with interrelated documents. The present chapter shows how to deal with the individual parts of the whole.

Who is the reader? In most cases the primary reader is the party on the other side of the contractual divide. It is best to maintain this approach of writing for the other side and not be overconcerned about the reader on the same side. Often technical specifications need to have a double identity: one is the technical support for funding and the other the instructions to the contractor. This dual approach is very unsatisfactory, because the two sets of readers are quite different and are seeking different things from the document. A specification that was contained within the financial submission to a board of directors on how to achieve a technical goal cannot as easily sit within a set of contract documents and function as a set of instructions.

To change from one function to another requires more from the specifier than simply changing the cover.

However, because of the work involved in producing a large specification, the specifier will often not have the time to rewrite for each audience in the chain. Therefore, it is best to identify the 'end-user' and write for that person alone. It is very much easier for specifiers to explain to the manager, the board of directors or the client that they have written a specification for the contractor to follow than for the client to explain to the contractor, after work has started, 'it was only written like that to satisfy the accountants on the board; we really didn't mean it'. The identification of the reader also applies if the specification is for further design – the detailed designer is the end-user.

The specifier must face a secondary question: has the reader the resources to comply? This problem will often govern how the specifier subdivides the specification to allow for specialists to carry out particular parts. For example, telecommunications contractors working with building and mechanical contractors will not perform well if what they have to supply is specified room by room or by individual items of equipment. The specifier needs a reader plan.

Finally, how to say it? This is the most difficult problem of all, and one which writers have to solve in their own way. Specifiers work with certain constraints. Although there will be a blank sheet of paper, there will not, hopefully, be a blank mind. When the specifier has answered the 'what?' and 'who?' questions, the problem of 'how?' may not loom so large, especially if the specifier is working to a plan. However, the specifier needs to recognize that most readers suffer from the 'I could have put it better' syndrome, for which the only known medication is for specifiers to demonstrate that they have thought out a plan and kept to it. If the medication is unsuccessful, surgery is usually necessary!

Beginnings, middles and ends

Specifications always require a beginning. Every chapter, paragraph, or sentence must begin somewhere. Illogical starting-points are the best way to confuse a reader. A beginning such as 'This section fully describes the Contractor's responsibilities with respect to certification...', followed two sentences later by description of some completely different responsibility, will immediately set a reader off on the wrong track and possibly lead to an unwelcome restriction of responsibilities. Similarly, beginning with 'This paragraph defines the terms Job Package, Work Package, and Work Breakdown Structure...' and then omitting one definition will lead to confusion. The logical beginning is the specifier's best friend because, in a few words, it both acts as a reminder of what to write and warns the reader what to expect. This is *not* the same as summarizing or duplicating what is to come.

The middle of a specification is an unsatisfactory place in many respects. There is no doubt that such a place exists, but the problem is where it ends. The beginning of the specification will have dealt with setting out the context, but the specifier still has to put the major part of the message across. The middle needs several subdivisions to keep it from becoming just a mass of verbiage.

In order to identify the middle, the specifier needs to have in mind the form of the ending. Unlike stories, specifications do not need to have endings. In practice, many specifications simply stop; however, this is not always good practice. The reader should be able to see that there is more to come on a subject, if there is more. The use of a paragraph such as: 'Related specifications: further details of the SCADA system appear in Sections 3.4, 5.8, and 5.10' signals to the reader that there is more, even if there are no cross-references in the text. It is important that the specifier heads the section with an appropriate heading because it will then be apparent in the contents section that there is related material and that the specification does not exist all in the one place.

Another use for the section or chapter end is to specify various miscellaneous items that cannot appear elsewhere because they do not require a section to themselves.

Sketching the plan

The blank sheet of paper is not always conducive to the act of writing. No computer manufacturer would call the blank vertical page 'user friendly'. As little can be done about the paper, something must be done about the approach to it.

The key is often to turn the piece of paper on its side and approach it from this new angle. The specifier should then divide it into three equal vertical sections and label them 'beginning', 'middle' and 'end' across the page (see Figure 5.1). The intention of this simple manoeuvre is not only to reduce the blankness of the sheet, but to provide the basic layout of the text. Now, starting anywhere on the sheet, the specifier should write down whatever comes into his or her head; it is not too important to put the ideas into the correct areas as the point of these initial jottings is to unscramble one's ideas. Ideas, then, will appear as a series of disjointed words, as in the illustration. Only when all the ideas have been exhausted should there be any attempt to put them into order, beyond placing them in the correct vertical column. Everything should now appear that is relevant to the text that will eventually be written, including notes on sources of information. Crossing out will be necessary in the process, but the specifier should keep this to a minimum at the preliminary stages.

The next action is to number each item in each column, showing its place in the final text. By now the specifier should have some conception of the layout of the whole text and what needs to be said. This is obviously a good

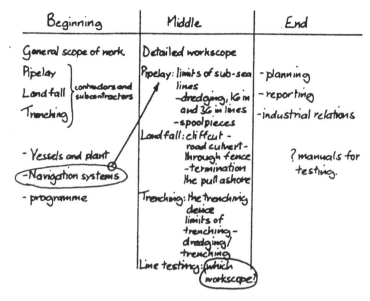

Figure 5.1 The preliminary plan for writing a specification

time to discuss the final text with colleagues in order to develop any further points that may have been missed.

First plan to first draft

After the specifier has completed the plan, as outlined in the previous section, writing starts in earnest. The best way, with the first draft, is 'to write in haste, correct at leisure'. There is no point in attempting to be word perfect the first time. Assuming that the specifier follows the plan that has been previously mapped out, the drafting will proceed in an orderly fashion with pauses only for matters of detail.

Once the draft has been finished, it is often best not to look at it again until the next day. A delay until the following day will probably not matter much in terms of production of the document, but specifiers will find that it will make a great deal of difference to their ability to read critically. When the draft has been reread, it is time to pass it to another for a check. (The advice that the specifier should ask a colleague to come in an hour early, sit in a locked room and read the whole draft is sound, if somewhat impracticable.)

Handling input from others

Few specifiers have the luxury or technical ability to write the whole of the specification; there is usually input from others that needs to be taken into account. The relationship between the originator of a document and others who will add to it is probably one of the most difficult to manage in the working world. It entails chasing, cajoling and changing what others have written. It is the last activity that causes the most problems. Document management systems, which are common in many organizations, are a useful aid in the allocation and control of documents written by multiple writers. Sections can be allocated to sole owners in the early stages of drafting, and these can be brought together into a single controlled document towards the end of the drafting process. Reading and writing access rights together with approval procedures can be set and adjusted as necessary to suit the document production process. A book on writing specifications cannot deal with the working relationship, but it is as well to remember that people become rather attached to what they write and react negatively to its defilement. Tact is the key.

The first task is to make sure that the formal definitions have been established and are used correctly and consistently. Thereafter, it is a matter of checking that the technical content matches in all sections – this can be achieved by specifiers discussing a common draft of headings and common terms which might need formal definition before drafting starts. Even this is no proof against inconsistency, and specifiers would do well to note as they prepare their drafts all areas of possible overlap with others and then resolve these before the final assembly of the specification. Of course, inconsistency does not apply only to repetition, but also to omission, which may occur because each specifier thinks that another has included an item or description. The omissions are harder to detect in the final assembled draft, if only because the person who undertakes the final editing may not be an expert in all necessary inputs.

The question of differing styles may involve more work in correction, though the specifier must judge whether the effort of editing to style is worth it. In the case of a minority input there is usually no need to alter drafting in order to conform to style. In any case, it is probably best to wait until the majority input is complete.

What must others write?

Specifiers must be in no doubt as to what to expect others to write. They spend time working out what they will write themselves, therefore it is important that others come up to scratch in their parts. The best way to obtain the correct input from others is to instruct them, and it is amazing how this self-evident truth escapes most specifiers. (It is quite common for specifiers to write other people's input for them, badly.) Specifiers should write down the items that they expect to be covered by others – and who the others are. When the first draft is issued, the specifier can then list those items that are expected to be covered elsewhere. It is important to bear in mind that receiving such a list would help others in planning their own work.

From the general to the particular

The specification is part of a journey: like any section of a journey the specification has length. The journey lasts the whole document and goes from the general to the particular.

The conditions of contract contain all the general directions that the parties agree should govern the conduct of the work. The journey then follows the hierarchy of the documents from the most general statements in the contract agreement, down to the drawings, which contain the most detailed description. However, the view of the documents is not as simple as it may at first appear. How the documents are seen depends on what the reader requires of them. The lawyer will see the conditions of contract as the starting-point on a matter which then opens out to a mass of description in the specification. For example, the lawyer will consider the drawings in the following light:

> The Contractor shall be liable for mistakes, errors and omissions in the drawings produced by him under the contract and for the adequacy of the information that they contain...

But the specifier will regard the drawings as a series of dimensions and tolerances, material lists and construction details that have to be translated into physical objects on site; the specifier is at the end of the journey from the general to the particular. In any dispute on what the drawings should contain the two readers will seek their reasoning in their respective parts. The specifier sees the detail of the specification as the narrowing down of that journey; the lawyer sees it as a widening, offering many possible routes. The particular wording of the specification is, as far as the specifier is concerned, the ultimate testing-ground on which compliance must be judged. In writing the text, specifiers must be clear on how particular they need to be on any one topic; the fine balance between reliance on the conditions of contract and the need for an exact description depends on the item concerned.

As it is not practicable to list those matters that should always be addressed in detail, it remains for the specifier to remember that as a matter of interpretation in most legal documents, 'the particular expression of an idea ousts the general reference to it'.

The specifier must also bear in mind that some, but not all, model forms of contract provide in the agreement for the hierarchy of documents within that contract. The IChemE forms are an example:

* The Agreement;
* The Special Conditions;
* The General Conditions;
* The Schedules;
* The Specification.

However, within each category the rule of the particular over-riding the general will apply.

Active and passive

When writing anything, there is a choice between the active and the passive voice as the means of expressing ideas. The growth of corporate culture in the modern world has been accompanied by the reluctance of individuals to take personal responsibility for what they write. Management is about doing the best for the corporation or the particular group that the individual represents. Whatever the cause of the anonymous presentation, it is achieved by use of the passive voice – no one ever 'does' anything, it is always 'done by' someone. Consider the following instruction on a printed envelope: 'Stamp shall be affixed by the sender', which is only a convoluted way of saying: 'Put a stamp here' (and much less elegant).

Although the use of the passive voice eliminates the personal pronoun 'I', it does not follow that the active introduces it. There is no need for specifications to become personal, and that is not the intention when using the active voice. The intention is to introduce greater clarity in the writing. Examine the following:

> The contract bank account shall be pre-funded by the Client in accordance with cash forecasts provided by the Contractor to meet invoice payments. Actual and anticipated cash requirements shall be forecast by month and amended at regular intervals.

Because of the use of the passive in the first sentence, the second loses its subject and it is therefore not clear who will provide the forecast of cash requirements. If the specifier had used the active in the first sentence, as in: 'The Contractor shall provide cash forecasts to meet invoice payments in accordance with which the Client shall pre-fund the contract bank account', then the second sentence could have used the same subject: 'The Contractor shall forecast the actual and anticipated cash requirements by month and amend them at regular intervals'.

The mistake in the above extract is very common, and no amount of logical argument can resolve the matter of *who* is to carry out the activity in the second sentence. Thus, the specifier has to be impersonal without sacrificing the need for clarity. Any piece of text may appear more 'immediate' in the active voice; this immediacy can help to elicit a positive response.

Fortunately, we now have tools for checking grammar in word-processing packages, which can be set to identify any use of the passive voice. Although this can be a little overpowering in normal use, when some use of the passive voice can vary the sense of the writing, it can be extremely helpful in specifications.

Flow diagrams and logic

In complicated clauses specifiers may find that development of the text starts getting out of hand. Reasoning purely by means of the written word is a process for which most technical people are untrained, and therefore the best solution may be to find a means that employs some form of technical input. The flow diagram is a well-known means. It should appear more often in the construction of logical argument. Specifiers might either use the flow diagram before they write the text, as a guide to the writing, or afterwards to check that the logic is sound. Consider the following situation:

> The client and contractor both have offices in the same town. The design and procurement contract is reimbursable. There are remote suppliers' offices. The client does not want to pay for travel time to the remote suppliers' offices, but is willing to pay for local travel time but not pay local travel costs, and pay all travel costs to the remote offices where the client has approved the journey beforehand. In some cases the client will pay for travel time to the remote suppliers' offices.

Although this might seem a reasonable description of the client's desires, it is no more than that. But if the specifier splits the problem into the two elements of cost and time, a flow chart can be built, as shown in Figure 5.2. From the flow chart the following clause almost drops out, noting that there are four solutions (time paid or not, costs paid or not):

The Client shall not pay for travelling time unless:

1   the journey is a local journey between the Contractor's own offices and the Client's office; or
2   the Client has given prior approval to the journey and its purpose; or
3   the Contractor has specific permission to travel in working hours.

The Client shall reimburse travel costs if:

1   the Client has given prior approval to the journey and its purpose; and
2   the journey is not a local journey between the Contractor's own offices and the Client's office.

The specifier can decide whether the clause is necessary at all, given that the flow diagram is self-explanatory. To use both flow diagram and clause is repetitive, and undesirable.

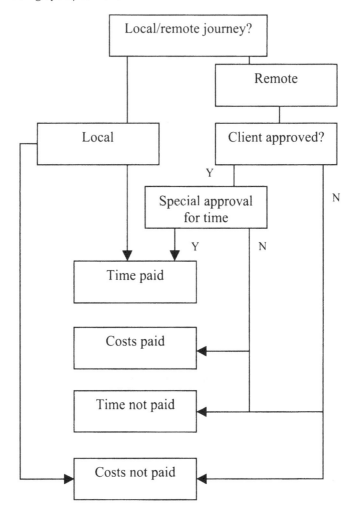

Figure 5.2 Flow chart for the payment of costs and time

Interlocking workscopes

The specifier has to accept the inevitability of more than one contractor at work, on a particular site, at the same time. Therefore, there will be interlocking workscopes; this will tend to cut across whatever the specifier has written. In the text the specifier must make clear what are the contractor's *interfaces* with others. The specifier can use simple phrases such as 'by others' as necessary to make the point that the contractor is not to carry out a particular task.

In the matter of actually defining the scope of work of each contractor, the flow diagram as an organizational chart for the project comes into its own. It becomes less of a chart of flows than a map delineating boundaries. In any case in which the specifier is faced with a complex set of workscopes, he or she should draw a map prior to setting any words down on paper. The example in Figure 5.3 shows the various jobs that contractors have to perform in the installation of a large water supply project – the type of scheme that was quite common in this country in the nineteenth and early twentieth centuries. Now such a scheme is more likely to occur overseas in areas of growing industrial development. It is undoubtedly true that the whole contract would have been simpler if one contractor had been given the whole task and subcontracted it at will, but this is rarely possible in large international projects. However, even in that case, although it may have simplified the contractual arrangements, the work description itself is not simplified. Using a map as in the illustration can help to explain the situation. The specifier, then, might give some thought to the inclusion of such a map in the tender documents, for if it helps to explain the work within the specifier's own organization it might help tenderers too.

Use of definitions

Why so simple a device as setting a definition is so underused is something of a mystery. Such underuse has two major consequences: a lack of clarity and overlong documents. Definition by assumption is hardly a proper substitute. In the conditions of contract are found a number of definitions, to which specifiers may have contributed, and which they must follow with precision throughout their own text. But from a purely technical point of view, specifiers require additional definitions.

The setting out of the particular meanings of words that the specifier intends to use is an integral part of planning the writing. Definitions, too, provide part of the content of the document. Specifiers, in setting out their stalls, find that they are disciplined to think in the way that readers think – and readers must continually ask themselves what is meant by the words they read.

The setting of definitions helps the specifier to tell, in addition to the final reader, others involved in the drafting what to follow. To do this, a list of 'defining terms' is required at the start of the sections in which they first occur. It can also be helpful to gather all the defined terms together at the beginning of each major section. However, the specifier must control the original, otherwise the definitions, which may undergo changes in subsequent sections, will appear without corrections in preceding parts. Constant reminders help to retain a discipline among the drafting team, which otherwise can be anarchic. Here, again, an electronic document management system is an invaluable tool in helping to maintain this discipline. The agreed definitions can be made available to the writers of each section and they need only be brought together in a single section of the final document.

Of course, the use of defined terms can save a large amount of time and

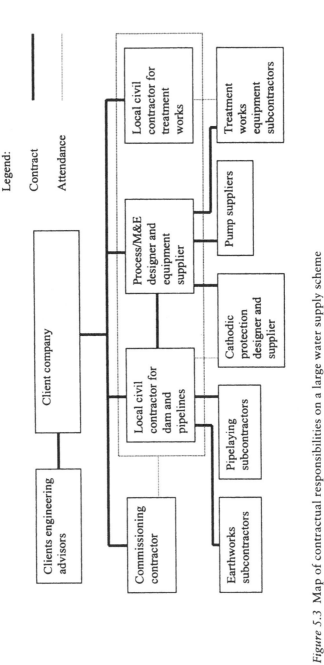

Figure 5.3 Map of contractual responsibilities on a large water supply scheme

paper because it generally means that the specifier can use one word instead of several. Consider the following:

> The Contractor shall confirm the layout of the control room within 15 working days of the commencement of the design. Working days do not include weekends.

The specifier is clearly on the right track in remembering that the contractor is unlikely to include weekends in the planning, even if some work is in fact carried out at weekends. However, the explanation also appears to be an afterthought and is incomplete. A proper definition would have included reference to bank holidays and public holidays, and possibly industry closedown periods as well.

The defined term is usually distinguished by having a capital letter, or being written in full capitals (making it less easy to miss). And the defined term does not have to follow the normal use or meaning of the word, for example:

> 'Ex Works' shall mean at the gate of the producing yard in the UK, and 'Ex Works Cost' shall mean the cost at the yard gate not including any head office overheads.

More usually, the specifier will be concerned with making a normal meaning absolutely clear, as here:

> 'As Built Drawings' shall be those drawings that record the state of the Plant at the end of the Maintenance Period.

The consequence of using defined terms with capital letters is that, if the specifier uses the same terms without capitals, they will be assumed to have different meanings to apparently identical terms with capitals. However, not all defined terms need have capitals, as here:

> ... 'pipe' shall always mean pipe to the tolerances in ASTM A530.

If, during writing, the specifier keeps to hand a list of words that are likely to have more than one meaning, these can be collected together at the end of the first draft to check whether or not it is necessary to define them formally.

Use of descriptive words and phrases

If specifiers are concerned with the use of definitions, they will also find that all rules are made to be broken. On occasions a particular matter cannot follow the general definition that has been carefully set. In such cases a sub-definition is used, or a descriptive word or phrase. Take, for example, the following as a definition:

> Working day shall be any day that the Contractor works but shall not include weekends or bank or public holidays.

If the specifier has defined 'working day' as above, difficulties may be encountered when it comes to the payment clauses because it will probably be the case that the client will pay for days worked whatever those days might be. At the relevant point the specifier can introduce a sub-definition:

> For the purpose of payment only, working day shall include all days actually worked by the Contractor.

Alternatively:

> In this clause Working Day shall include bank and public holidays if worked.

Headings and definitions

There is no convention which says that headings are a substitute for definitions, to use a word or words in a heading is not effective as a definition. In fact many standard conditions of contract state the following, or something similar:

> The headings and marginal notes shall not be used in the interpretation of the Contract.

Interpretation of lists

It is a fact that lists are very difficult to write unless everything that the specifiers intend is included without exception. The difficulty is that if the specifier omits anything from the list, then it will be assumed that it is omitted on purpose. And it is not safe to assume that the specifier can overcome the problem by the insertion of any general words to sweep up those items omitted. There is a rule in drafting, so that:

> ... general words used at the end of a list of items of a particular kind take their meaning from the same kind as the items on the list.

Therefore, under the general rule, if the specifier writes the following clause, it will not be possible to prevent the contractor painting the wood if, for example, the moisture content is too high:

> The Contractor shall ensure that surfaces for painting are free from warps, cracks, shakes, holidays or any other defect whatsoever.

Moisture content may well be critical to painting, for if it is too high then the paint will not bind. However, if it is not mentioned, can it be inferred? The answer is probably that it cannot because 'warps, cracks, etc.' are defects and moisture content is a state or natural property of wood. And the words 'or any other defect whatsoever' are of no use because of the rule that the list cannot be expanded to include other items unless those other items can be deduced logically from the list following the same train of thought. Like all such legal rules it has a Latin name (*ejusdem generis*), which is given here only as a point of information, in case the specifier finds that only the Latin name has been used in the particular conditions in a contract to the effect that 'the ejusdem generis rule shall not apply'. It does happen and it can be used to alter a fundamental principle in an almost clandestine way. If specifiers want to escape the effect of the rule, they can use a phrase such as 'whether listed in this paragraph or not' at the end of the list.

The problem also occurs in the reverse situation, in which a general statement precedes the list, such as:

> The Contractor's equipment shall be generally capable of operating in the hostile environment on the Site, and be properly serviced, clean, mechanically sound and fully equipped with the necessary attachments.

The client cannot enforce under the clause a requirement that the equipment must be new, on the basis that the client considers that only new equipment can cope with the conditions. A specifier who wants to overcome the constraints of a list should introduce a 'sweeper' phrase at the beginning of the list, as here:

> The Contractor's equipment shall be generally capable of operating in the hostile environment on the Site, and *(for example, but not exclusively)* be properly serviced, clean, mechanically sound and fully equipped with all the necessary attachments.

Even so, it is unlikely that such a radical demand as for new equipment could be enforced under such wording. If new is required, the specifier should say so. The specifier should also note that the use of 'etc.' would be almost useless in expanding a list.

Paragraph structure

In order for specifiers to get their message across, they need to think of the optimum structure of the paragraphs that are used. In paragraphs which are intended to set out a chain of events an orderly approach is necessary.

The first element is the 'reason why' a particular event has to take place, or the circumstance in which the event is likely to occur, for example:

> This section shall only apply if the accommodation module and helideck is a separate detailed, design, procure, and fabricate contract.

The next stage is the intention of the parties:

> The Contractor's responsibilities will be modified to the extent shown in this Section.

The specifier then needs to show under what conditions the parties will carry out the intentions, as in:

> The Contractor shall carry out preliminary design studies and develop final optimized arrangements in accordance with the design criteria given in the Technical Specification.

Finally, the specifier shows the exceptions:

> The Contractor shall not, where this option applies, provide the services described in Section 2.9 for the fabrication, installation, hook-up and commissioning of the accommodation module.

Through the four stages the specifier has covered all the elements of a particular chain of events and has, by the use of the four introductory paragraphs, set down the approach to the whole section.

Types of sentence

Sentences exist in four types: imperative, suggestive, explanatory and precautionary. The specifier's use of any one of them depends on the responsibility that the contractor is to bear for any one action or item. Examples of the four types are as follows:

- *Imperative:* The Contractor shall carry out the preliminary design.
- *Suggestive:* The Contractor may propose an alternative strategy for the accommodation module.
- *Explanatory:* The Client expects a closer involvement with the Contractor for the accommodation module design than for the other parts of the topsides design.
- *Precautionary:* The Contractor shall ensure that all interfaces between the accommodation module and the topsides are compatible, and that discrepancies are advised to the Client.

Provided that the specifier keeps the type of sentence clearly in mind, the use of the correct wording should follow without difficulty.

Contents list

By the time a large chunk of text has been completed, the specifier has probably forgotten what is in it. In any case, the text will have gone through a number of drafts and, as a result, it may have finally come to rest out of its originally planned order. To check whether this is the case, the specifier should write the contents list, or generate the contents list using the appropriate tools on the word-processing software, using the paragraph or section headings. The act of writing the contents list is a good reminder of some comment or idea that may have been lost along the way.

Cross-references

The specifier should start from the position that cross-references are best avoided. Sometimes text becomes impossible to read because of the proliferation of numbers referring back and forth. If the need for cross-references is growing out of proportion, i.e. more than one or two cross-references every two pages of text, then planning a new layout should be considered. However, it is often the case that cross-references are helpful in an early draft of a document. The use of cross-references at the early stage reassures readers (the writers of other sections in the specification) that proper details do appear elsewhere, and the specifier can simply score them out in later drafts without loss of meaning.

The difficulty with cross-references is that they lead to errors and confusion as the specifier emends the draft and particular clause numbers. Quite often, a cross-reference, such as 'see Clause 3.15,' which was intended to be about one thing, e.g. starter motors, might actually refer to a clause dealing with something else entirely, e.g. cables. The best way to avoid this problem is to make the cross-reference more general by referring to a clearly identifiable section of the specification rather than to a precise sub-clause. This defeats the purpose of the cross-reference to some extent, but it will not be too onerous for the reader, provided that the section is not more than four or five pages long, and it will act as a useful prompt to look elsewhere in the document.

Accuracy and clarity in writing specifications

Verbosity

Verbosity does not usually arise because specifiers actively seek it, but rather because they are unsure of the audience or lack a writing plan. Being unaware of the reader's knowledge may encourage a specifier to write down everything possible on a given subject. Too much explanation in a specification is more likely to be a fault than too little. The contractor needs to know sufficient and no more. The contractor is an expert in the field.

Another source of verbosity is the dictated specification – even a dictated memo is hard to control. The specifier will do well to avoid dictation as a means of drafting. One frequent manifestation of verbosity may be referred to as the 'this is because' syndrome, for example:

> The Contractor shall provide the monthly report within five working days of the end of the month. This is because the Client has to consolidate the report into an activity report for its London office by the 19th of the month.

The second part of the clause is relevant to the client but irrelevant to the contractor; it only adds words. The specifier should consider the question: 'if the contractor does not know this, will it affect his or her obligations under the contract?' If the answer is 'no', then the unnecessary words should be omitted from the specification. However, the second sentence in the example above might be a suitable addition to an annotated clause for use in explanation to the client's staff on site.

Repetition and duplication

If there were a league of heinous drafting crimes, repetition and duplication would be near the top. All repetition runs the risk of imperfect reproduction of an instruction, and therefore can add to the likelihood of a dispute. If a subject appears more than once in a document, the specifier may well fail to alter one of the occurrences of the wording emended during the editing process. Correction in one instance of 'will' for 'shall' (see Chapter 6) might make for a dispute. For example:

> The Contractor will set up a quality review team for the duration of the contract.

This original wording of the subject, after second thoughts on the specifier's part, may have become:

> The Contractor shall set up a quality review team for the duration of the contract.

If both versions appeared in the contract, the contractor might argue, on disbanding the team before the end of the contract, that the use of 'will' left it to his discretion, the later amendment to 'shall' being in a lower document in the hierarchy and therefore not binding on the contractor. The specifier should refrain from repetition 'to make it clear' – it rarely does more than confuse.

Length

The length of a piece of writing is not an absolute guide to its appropriateness. Many readers may feel that whatever the length of a document, it could have been less! Specifiers, it seems, are fairly consistent in the length of documents they produce, in comparative terms. Length is also dictated, to some extent, by layout, and there is the concept of apparent length to contend with too, because some layouts make a document appear longer than it actually is.

Avoiding 'legalese'

The specifier must avoid drafting as if he or she were a lawyer. No doubt, this is an unwarranted slur on the legal profession because most specifications would be improved if lawyers had a hand in drafting them. However, specifiers are not usually lawyers, and therefore should not employ what they perceive as legal construction or phraseology.

But what is 'legalese' and how to avoid it? Consider the following:

> The Contractor shall not except as aforesaid in the foregoing paragraph allow such nuisance to be committed by the Subcontractor as to cause the said Client to suffer loss damage inconvenience or delay (whether attributable or not) in the installation of the said Client's roof.

The paragraph is plainly incomprehensible not only because the specifier has not planned what to say, but also because it has been based on a sixteenth-century lease and omits all the punctuation. The use of words such as 'aforesaid', 'nuisance', 'said', and so on, is quite unnecessary and out of place. The paragraph is much better as:

> The Contractor shall not allow the Subcontractor to interfere in the construction of the Client's roof.

As soon as the specifier feels that sentences are becoming tortuous, he or she should be on guard.

Avoiding the obvious

Specifiers are writing for an expert readership, and therefore a statement like the following represents a waste of words:

> The embankments shall be watertight and shall show no signs of leakage or seepage or other form of water egress.

If the embankment is to be watertight, then it must not leak. The contractor does not need to be told any more. Similarly:

> The Contractor shall ensure that his workmen use crawling boards on the corrugated roofs as they are in a weakened condition and will not support the weight of a man. Serious injury could result if the Contractor ignores this instruction.

This goes too far in warning an experienced contractor. Also, the addition of words may affect the meaning of the paragraph. In the above example, for instance, it could affect the ultimate liability for damage to the roof itself because the second sentence refers only to personal injury.

Switching mood

When the specifier has settled on the mood of the writing, that is in the context of the active and passive voices, the imperative and suggestive styles, and the personal and impersonal forms of address, this must be consistently maintained throughout all the subsequent drafts. It is very easy to slip from the active to the passive voice, even within the same sentence, as in:

> The Contractor shall comply with BS 5432 and adequate quality assurance procedures shall be set up by him.

Though this may be untidy drafting, it is not usually hazardous if the switch occurs in the same sentence (but see comments on the use of the passive voice on p. 92). A switch from the imperative to the suggestive, or vice versa, is more serious. Take, for example, the following two extracts:

> (1) ... may approve the use of the Contractor's procedures...

> (2) ... shall approve the Contractor's procedures...

If (1) becomes (2), then the client has no option in the second case but to approve whatever the contractor submits to him. The specifier can never have intended that to happen.

It would be unusual for the specifier to use the third-person singular or plural pronouns in the drafting, but there would be no great risk. However, if, in the context of referring to the contractor's agent, there were to be a change of reference back to the contractor, it would have the effect of making the performance of the obligation impossible by the agent personally. To avoid such risks, the specifier should list in a general section the titles of those who have specific responsibilities under the contract, and not allow loose references to confuse the issue.

Change of language equals change of meaning

The golden rule is: do not change the language, unless a change of meaning is intended. Change in the description is sometimes referred to as 'elegant variation', and it is a widely used technique in writing literature, but it can lead to problems in writing specifications. Definitions and implied definitions must be adhered to. To start off by calling the contractor the 'main contractor' and then change to the 'purchasing contractor' presents no great problem, but it would indeed be confusing if there were two contractors, only one of whom was responsible for purchasing. The specifier should not abhor the continual use of the same word or phrase for the same action or item.

Vagueness

The purpose of specifications is to make something clear, so that action will follow. There is no excuse for vagueness of expression. To put it as shown below does not give the contractor much hope that time invested will be worthwhile:

> The Client might consider proposals by the Contractor for…

On the other hand, the following phrase leaves the decision whether to make a proposal or not open to the contractor:

> The Contractor may propose for the Client's approval…

If options are offered, then the specifier should not make one option dependent on another.

Pious intentions

Phrases setting out vain hopes that something may be done by someone have no place in the documents, for example:

> The Client hopes to make available to the Contractor…

This can be taken to the point at which a long description of a whole project comes into the class of an unrealizable hope. There is also the danger that the contractor will treat such hopes as warranties that a particular state of affairs will exist. The contractor may have a valid claim if it can be shown that he or she relied on a warranty.

Words of similar appearance

Specifiers may make life difficult for themselves if they use words of similar appearance when one of those words is a defined term. For example, if

'Works' is a defined term, the specifier should avoid referring to the 'work' to be completed by the contractor. If the specifier does not avoid the use of similar words in this way, it will soon happen that a word will be used incorrectly, for example:

> The 'Works' shall mean the pump station and all its appurtenances as shown on the Drawings. The Contractor shall carry out the works in such a way...

Confusion will arise if 'works' is accidentally typed as 'Works', for clearly 'Works' is an entity, whereas the second reference is to an activity. It would have been much better for the specifier to have used 'design' or 'construction' as necessary. It is even worse to define two words whose similarity can be foreseen to lead to confusion, for example:

> 'Line Pipe' shall mean the 36-in. or 16-in. coated pipe stored at the pipecoater's yard for collection and loading out by Contractor.

> 'Pipe' shall mean the 36-in. or 16-in. uncoated pipe.

> 'Pipeline' shall mean the 36-in. or 16-in. pipe as laid on the sea-bed.

Too many adjectives?

In a specification adjectives are not really necessary, unless they form part of a set definition or can be inferred from common usage in the industry to describe a state; for example:

> Surfaces to be painted shall be free from dirt and be in a clean, sound, workmanlike condition.

This introduces the adjective 'workmanlike', which has no precise meaning. On the other hand, an adjective, such as 'approved', for which the conditions of contract have set the definition can be used correctly as follows:

> 'Approved' shall mean approval by the client in writing that the specification has been complied with.

Similarly, the use of the word 'wet' in the context of 'wet gas' is correct as it is a technical term describing a gas with certain chemical properties. But 'wet weather' is more doubtful as the degree of wetness is probably important:

> The Contractor shall not apply paint in wet weather.

The statement above really does not address the relationship between wetness and the inability of paint to adhere to wood or metal surfaces. It would have been better to write:

> The Contractor shall not apply paint when the air humidity is greater than 90% or when moisture is visibly present on surfaces to be painted.

Sentence length

In addition to what appears in Chapter 6 on sentences, there is the matter of the optimum length of sentences, and that has to do with readability. Long sentences are difficult to follow, and they have become less and less acceptable in documents. In fact, the optimum length of a single sentence is supposed to be around thirteen words; it seems that number comes from newspapers, in which journalists have to put over their material in a confined space. However, the comparison with journalism may be spurious in a book about writing engineering specifications if only for the fact that journalists sometimes have no great spur to get their facts correct in detail.

Nonetheless, specifiers should pay attention to the length of their sentences and always try to reduce them. Sentences should contain a single concept, with no more than one exception, and not switch subjects. The following example contains these faults:

> Road surfaces shall be to line and level with a maximum vertical tolerance of 15 mm in 100 m and road drain inverts the same, except that where road drains meet field drains where the tolerance shall be that of the field drain, and underpass height shall never be less than 4 m.

In a sentence in which there are two concepts – the tolerance of roads and drains – and two exceptions – the field drain junction tolerance and the underpass tolerance – and two subjects – the road and the underpass – it would have been far better to have written:

> Road surfaces shall be to line and level with a maximum vertical tolerance of 15 mm in 100 m. Road drain inverts shall be to line and level with a maximum vertical tolerance of 15 mm in 100 m, except where road drains meet field drains where the vertical tolerance shall be that of the latter. Vertical tolerance can be increased in underpasses to maintain a vertical height of 4 m minimum.

Two words where one will do

Specifiers may be assailed by doubts when describing some act or state of affairs. This can lead to elaborate construction or description along the lines of the following:

> If the goods supplied arrive at Site having suffered loss, damage, breakage, or in any other way be incomplete the Client shall not be required to accept them.

There are simply too many words to describe the fact that the goods must arrive in accordance with the specification. The specifier seems to have lost faith in previous wording, and should have been able to say:

> The Client will not accept goods at Site unless they accord with the Specification.

Even if specifiers use several words which are not apparently covered by what has already been written, they need to ask themselves whether they have added anything to the original meaning, as in the following:

> The Contractor shall carry out the directions, instructions, or orders of the Client's Manager at Site.

The use of synonyms for 'instructions' has added nothing to the meaning.

It is sometimes the case that the use of several words is forced on the specifier because of an earlier use of words. If the document has the words 'orders, instructions and directions' all issued by the 'Manager' the specifier should tackle the problem at source, i.e. set a definition and use a single word thereafter.

Use of technical words

The specifier has to use words in their correct place. This applies to technical words, but precise meanings must be maintained, even when non-technical people read the text. This precludes the specifier from using jargon, though it is often true that today's jargon is tomorrow's technical usage. Technical words are the key to concise drafting, and specifications are not meant for the lay reader. But the specifier should not assume that meanings are fixed, and the status of words in this respect should be queried.

To overcome the problem of floating meanings, the specifier must ensure that words not in daily use are referenced to some recognized standard in which the meaning of technical terms is not set by definition. A glossary is particularly useful for abbreviations.

Use of acronyms

Strictly speaking, acronyms are words formed from the initial letters of words in a description. In engineering, and in other fields, the need to create a word that can be spoken is not rigidly held, and what is often an abbreviation of a technical phrase is referred to as an acronym. The difference between an acronym and an abbreviation (see p. 143) is not always clear in the technical world. Three-letter acronyms (or TLAs to the initiated) are not the same as abbreviations. Some acronyms have more than three letters, but, surprisingly, the majority do have three letters. They are a special type of abbreviation that is used as a code for technical words and jargon. Acronyms are always made up from the initial letters of the composite words, but abbreviations often include more than the initial letters, for example 'IChemE' – the abbreviation for the Institution of Chemical Engineers. Acronyms serve the same purpose in specifications as abbreviations – they do make the writing more concise and easier to follow and it is much easier to write 'PLC' in a specification for instrumentation, control and automation (ICA) than it is to write 'programmable logic controllers'. The majority of readers of the specification will know what is meant without any explanation, but some may not, and to make the document as easy to read as possible and to make it accessible outside the 'brotherhood' of specialists, the specifier should include all acronyms in the glossary.

Expressions of scope

When specifying the scope of the contractor's work, there is a tendency to develop 'comfort phrases' to cover up the difficulty of detailing exactly what the contractor has to do. If the specification says 'The Contractor shall, without limitation, carry out...', then it may well list a number of items and leave the phrase 'without limitation' to cover all that cannot be defined. For the specifier to use such a sweeper clause in a reimbursable contract may be satisfactory, but how is the contractor to cope with the implied meaning in a fixed-price contract? Similarly, if specifiers hold the contractor responsible for 'among other things the custody of the Plant...', this begs the question of 'what other things?'

Phrases that indicate extensions of scope are suitable as a proper extension of the contractor's undefined responsibilities, but only if the other parts of the contract (usually the payment terms) allow such leeway.

Expressions of time

There are few commoner causes of dispute than those related to time. When specifiers make some statement of time, they are making a judgement about two things: that the required action can be performed in the time and that this period can be accurately specified; many documents however fail on both counts.

It is worth setting out common phrases used in specifications and what they mean: some examples are given in the following list:

- 'on' means the day is included; if a period of time is to begin on a *named* day, that day is included;

- 'on or about' means within a day or two either way;
- 'from' a day or the occurrence of some event has no clear or accepted meaning and the specifier should use another construction; instead of 'from the date of delivery of the pumps' write 'commencing on the date…';
- 'till' or 'until' is not well defined, it is better to say 'the period ending on the [date]';
- 'within' can be quite vague if there is no reference point. However, 'within seven [or some other period] days from the receipt of' is a common phrase in most standard conditions of contract. The specifier has the alternatives of 'by the end of the seventh day' or the 'contractor has seven days';
- 'day' can be midnight to midnight or from any hour to 24 hours later; it is better to say when the period is to start;
- 'clear day' means a period of 24 hours starting from midnight;
- 'month' is an odd one as it is defined by law to mean calendar month, going from date to date, or to the end of the month succeeding if there is no matching date;
- 'year' means a period lasting 365 days or 366 days if the period includes 29 February;
- reasonable time' is a common and unavoidable phrase which has many uses, but its exact meaning is always open to interpretation in a particular situation; try to use the exact period if possible;
- 'forthwith' and 'immediately' mean that whatever is to happen must do so without delay – there is no ambiguity. However, both can be difficult to achieve in most cases and the specifier may be creating an unnecessary contractual hurdle;
- 'as soon as possible' means just that, the possibility is the governing factor, not the time limit that one party puts on it as an interpretation.

Use of Latin

There are many good reasons for lawyers to use Latin phrases in writing precise legal documents. There are also many good reasons why they should not use Latin, particularly in contracts that involve the general consumer. The reason for not using this dead language in specifications is simply that it will not be understood by the majority of readers, including those abroad.

Use of 'and' and 'or'

If the specifier is careless over his or her use of the conjunctions 'and' and 'or', the consequences can be serious. It is quite clear that 'and' will mean 'together with' and not imply any sort of choice. For example:

The Contractor shall report manhours by discipline and system.

This means that the report must contain both discipline *and* system, whereas the following means that the contractor need only report against one:

> The Contractor shall report manhours by discipline or system.

It becomes more complicated if the specifier writes:

> The Contractor shall report manhours by location and discipline or system.

Now the choice becomes by 'location and discipline' or 'system', but not 'location *and* system'; unhappily, perhaps, the specifier wanted the last option.

If specifiers use a string of conjunctions, they must realize that these react with one another. It is usually clearer to resort to a list, as in:

> The Contractor shall report by:

- location and discipline; or
- location and system.

Use of 'and/or'

For some reason, the phrase 'and/or' rolls off the pen with great ease! For this reason, specifiers are in danger of using the construction in the wrong place. Take the following:

> The Contractor shall carry out the welding, inspection and testing of the tubular transition sections and/or the tubular intersections as required by the Purchaser.

If specifiers write this, then an alternative scope of work is being set out, i.e. '... the tubular transition sections and the tubular intersections, or one of them'. And if they write the following, then they are being illogical:

> The Contractor shall supply and/or fabricate the joints and locking wheels.

For, although the contractor can 'supply or fabricate', why should he 'supply *and* fabricate'? Here, the use of the word 'and' is superfluous. Similarly, in the following extract, the use of 'or' is superfluous:

> All welding undertaken in the assembly of the buoyancy and/or flooding control points shall...

In addition to these incorrect uses of 'and/or', the specifier should consider the burden to readers of the text in unravelling the exact meaning of this construction. The specifier should use it sparingly, if at all.

Use of 'except'

The exception or proviso is a means by which the specifier can modify a previous statement, for example:

> The Contractor shall pay for all the tests to the Plant except those which demonstrate that the Plant will operate in accordance with the specification.

However, specifiers frequently fall into a sort of 'legalese' and become entangled in phrases such as 'provided always that'; such constructions should be avoided.

Use of 'less than', 'more than', 'equal'

If the specifier writes 'less than 5 mm', then 5 mm and more is obviously excluded. If 'more than 5 mm' is written, 5 mm and less is similarly excluded. If the specifier writes equal to 5 mm', then no other value will do. This may seem rather obvious until one reads:

> The allowable tolerance shall be equal to 5 mm.

A tolerance should not be qualified in this way.

Numerical accuracy

When using numbers to describe a process rather than to put a dimension to a process, the specifier must be aware of slipping into the absurd, as here:

> Weld preparation fusion faces shall be 100% visually examined for defects.

The specifier should have kept to a more literal description to avoid the problem of measuring the percentage examination to see whether the contractor was complying with the contract.

Use of 'including' and 'excluding'

When specifiers are writing a list of tasks, or anything else that is likely to have exceptions, care is needed in stating what is included and what is excluded. Examine the following:

The Contractor shall report on his scope of services including planning, estimating, progress reporting, cost control, excluding procurement, fabrication, installation, and certification by using the Work Package system.

It is not clear in the above where the exclusion ends: are the services after 'excluding' all omitted or just the procurement? In fact, the specifier wished to exclude only the procurement, and therefore should have written:

The Contractor shall report on his scope of services including planning, estimating, progress reporting, cost control, fabrication, installation, and certification but not procurement, by using the Work Package system.

Use of 'including' and 'comprising'

There is a significant difference between 'including' and 'comprising'. When the specifier uses the word 'including' in a list, it is taken to mean that the items referenced are included but it leaves open the possibility that there may be others, for example:

The Contractor shall carry out soil tests including triaxial, shear vane, CBR, and moisture content tests.

However, the use of 'comprising' would indicate that the named items are included but that there are no other tests, for example:

The Contractor shall carry out soil tests comprising triaxial, shear vane, CBR, and moisture content tests.

By using the word 'comprising', the implication is stronger that nothing has been omitted.

Use of 'responsible for'

The phrase 'responsible for' has no meaning if there is no qualification of the extent of the responsibility or to whom it is owed, and for what the responsibility is owed. Examine the following requirement:

The Contractor is responsible for all tests carried out on materials supplied by him and incorporated into the Plant.

By using the phrase like this, the specifier will be unable to pursue defective materials satisfactorily, and an important chance of retesting or rectification may be lost. However, if the specifier qualifies the phrase, the resulting obligation becomes clearer (depending on what was originally intended) in the case in which the contractor has only to have tests carried out:

> The Contractor shall be responsible for ensuring that tests are carried out on...

If the contractor is responsible only for costs, the specifier should write:

> The Contractor shall be responsible for the costs of the tests carried out...

If it is intended that the contractor must carry out the tests and be responsible for the costs, the specifier should write:

> The Contractor shall be responsible for ensuring that tests are carried out at his cost on all materials supplied by him and incorporated into the Plant.

Use of 'e.g.' and 'i.e.'

Both abbreviations 'e.g.' and 'i.e.' are from the Latin, and have quite different meanings: 'e.g.' means 'for example' and 'i.e.' means 'that is'; confusion between them leads to trouble. If the specifier writes the following, then one particular type of compression damage to the exclusion of all others is specified:

> The Contractor shall not allow compression damage (i.e. buckling) in temporary members.

It should have been written:

> The Contractor shall not allow compression damage (e.g. buckling) in temporary members.

(See also problems with lists, p. 98)

Use of 'effect' and 'affect'

Both words have distinct meanings and they occur quite regularly in specifications. Affect has a meaning only as a verb, but effect has a meaning as a verb and as a noun; it usually occurs in specifications as a noun. The most common meaning of the verb affect is to alter some person or thing in some way or to have an effect on he, she or it. The verb affect is also used to mean pretend but this would rarely, if ever, be used in a specification. To effect has the completely different meaning of to bring about or to accomplish. An effect is, as we have seen, what happens when some person or thing is affected.

Elegant style

Style cannot be taught or imposed, but will necessarily be described as good/bad/elegant/clumsy/precise/muddled by readers and users of specifications. Good style comes out of the writing itself, growing with the writer, and there is no reason why specifiers cannot develop style in the way that they present the subject. There are several publications that deal with the correct writing of English. The specifier can decide which one he or she prefers as a reference.[1]

If style is elegance of expression, then the specifier should make a conscious effort to achieve it. Although it may arise out of continual writing, style does not come by accident. The basic ingredient is the plan of the writing; without a plan, it cannot grow because the specifier has to spend too long organizing ideas 'on the run' to have time to devote to style.

How, then, to achieve an elegant style? It can be achieved by being just a little bit different within the rules. Additionally, a relaxed 'conversational' approach to the initial draft can be the basis of a writing style. If specifiers say 'rules are not for me', then it is unlikely that they will be able to defend their style against those who do not approve. The business world expects people to seek approval for what they do and style, too, will be subject to approval. Therefore, it needs to be a conscious development that the specifier can defend.

As the specifier is concerned, above all, with accuracy, this is the first target. Beyond that, the aim should be to use sufficient words and no more to express ideas. The specifier must avoid jolting the reader, the text giving the feel of leading the reader on to further understanding, remembering that the reader is reading not for pleasure but for a purpose and is therefore not inclined to experiment with novel forms of written communication.

The house style of the specifier's employer may become a problem in some cases. But the specifier needs to understand that house style is no more than a set of rules to write *within*, not a constraint on writing, and the creative activity can still remain.

As a guide to style, the specifier should learn to recognize it in the work of others. This is not only a valuable part of the learning process, but a useful point to remember when editing other people's text; the art of précis is useful in this respect, distilling the writing of others and retaining its essential points. Minutes of meetings are also good practice – although often considered to be tiresome, these can be developed to a high degree of clarity and brevity, despite the ramblings that may occur in the meetings that they record.

Words and confusion: the Fog Index

Specifiers can attempt to establish objectively the clarity of their writing with the help of the Fog Index – so called because 'fog' is a measure of obscurity. In calculating the Fog Index (FI), the specifier proceeds by:

1   choosing, at random, a number of consecutive sentences containing in total approximately 100 words;
2   noting the number of sentences contained by the 100 words and calculating the average number of words per sentence (NWPS);
3   using the same 100 words, counting the number of words which contain three or more syllables, ignoring proper nouns and those words which are three syllables long because of '-ed' or '-es' (TSW);
4   adding NWPS to TSW and calculating 40% of the total.

Thus, the answer is the FI for a particular piece of writing. The lower the Fl, the more readable the work. The specifier should aim for an FI of twelve or less when writing non-technical text (many technical words are multisyllabic). If the FI is greater than twelve, the specifier should either shorten the sentences or use simpler words, or both.

Other 'readability' indexes are available in word-processing software, and specifiers may find alternative approaches which work well for them. However, some of these software tools are based on a different version of the language, namely US, rather than UK, English, and they may constrain the specifier to a style that they do not find natural.

6 Grammar

Alice felt dreadfully puzzled. The Hatter's remark seemed to have no sort of meaning in it, and yet it was certainly English.

Why grammar?

If an excuse is needed for including a chapter on grammar in a book on writing specifications, then it must be a strange, esoteric world that writers of technical documents inhabit! Yet it would appear from many documents that such isolation is encroaching. The arguments for and against the strict application of grammatical rules are many. In these days of text messages, which have evolved their own language and code, an absence of correct grammar in a written specification would hardly appear to be a major problem so long as the text can be understood. However, the point is that the specification should not only be understood but also be interpreted consistently by different readers. The first rule of interpretation of legal documents is to take the literal meaning of the words, and that means that the specification should follow the rules of grammar. Literary education for anyone opting for sciences at school often comes to a halt at the age of 16 or earlier. Recently the situation has improved, but access to basic reminders of writing grammatically correct English is strangely difficult, although there are many good publications on style and clear writing which do not address grammar directly.

The rules of grammar set out in this chapter are necessarily abbreviated to provide a basic guideline and to show how rules may be adapted to the situation. Technical writing and the employment of grammatical rules must be influenced both by the legal context, tending to extreme over- or underuse of punctuation, and by the broken layout of text involving tables, formulae, lists and explanation. No specification can be written in a flowing style like a novel, and therefore the rules of grammar that may be taught as aids to achieving a literary style must be re-examined.

In the end, it is the specifier's task to seek clarity, and by revising the basic grammatical rules it should be possible to progress through this chapter to the final layout of the text. An exploration of English grammar only starts here and, as in Chapter 1 on law, there are many ways forward.

The eight parts of speech

It is necessary in this chapter, as in all good specifications, to define the terms that will appear; this has the dual role of definition and revision.

There are eight parts of speech – the noun, verb, pronoun, adjective, adverb, preposition, conjunction and the interjection:

- *Nouns* are words that take the action, or have the action taken upon them, naming words.
- *Verbs* denote action, or describe a state of being.
- *Pronouns* stand in place of a noun.
- *Adjectives* modify a noun.
- *Adverbs* modify verbs, other adverbs and adjectives but never nouns.
- *Prepositions* introduce nouns and pronouns.
- *Conjunctions* act as connectors to join words, phrases, clauses and sentences.
- *Interjections* express a sudden emotion.

The important point to keep in mind is that words perform functions, and it is the particular function of a word that determines its classification.

The eight parts of speech are set out, together with their sub-parts, in Table 6.1. In setting out the examples, the words are shown in categories, but such a table cannot explain that a word is what it does. For example, the adjective strong' can be a noun:

Out of the strong came forth sweetness.

Similarly, the noun 'steel' can be an adjective:

Steel beams and columns shall be painted.

Verbs can lead more than one life in this respect because they are so flexible, as will later become apparent.

Subject and predicate

The sentence is the basic building-block of all writing – a group of words that makes complete sense on its own. However long or short it may be, the sentence contains two basic elements: the *subject* and the *predicate*.

The subject is the idea of the sentence and the part that carries out the action. The subject is either a noun or a pronoun, or derivatives or forms of either. The predicate is the name given to the part of the sentence that tells about the subject. Therefore, the predicate always contains the verb – and is often more complicated than the subject, especially in technical writing. So the form of the simple sentence is subject/predicate, as in this example:

Table 6.1 The eight parts of speech

Part of speech	Sub-part	Comment	Example
Noun	Collective Proper Common Abstract		The company, workforce, personnel London, Temporary Works, Contractor Steel, concrete, contractors Influence, suggestion, liability
Verb (active and passive)	Transitive Intransitive	Takes a direct object Does not have an object	To place (something), to instruct (someone) To succeed, to benefit
Pronoun	Subject Object		I, you, he, she, we, they Me, you, him, her, it, us, them
Adjective			Strong, approved, deciding
Adverb	Manner Time Place Degree	Says 'how' an action was done Says 'when' an action was done Says 'where' an action was done Answers the question 'to what extent?'	Quickly, tightly Early, late Along, down Very, rather, extremely, too
Preposition	Simple Compound	Always positioned in front of a noun or pronoun Written as a phrase, sometimes called a prepositional phrase	Beyond, in, to, down In accordance with, on or about
Conjunction	Single Paired	Join parts of a sentence which contain (though not always) their own verbs; when used in pairs, must be put before the words joined	Or, and, but Either... or, neither... nor
Interjection			Oops!, D'oh!, Ay Karumba!

The roads/shall be graded to a smooth finish.

The use of the active or the passive voice does not affect the form of the sentence.

There is no rule that the subject has to precede the predicate, though it is hard to construct a sentence that illustrates the reversal. Reversal is effective in literature, as here in the title of a book: *Quiet Flows the Don*. The word 'quiet' is an adverb and so cannot be the subject of the verb 'flows'. The subject is 'the Don', a reference to a river.

However, much more common is the splitting of the predicate to lie each side of the subject, that is part predicate/subject/part predicate, as in this example:

Prior to the commencement of painting/the Contractor/shall obtain the approval of the Engineer.

The above are all examples of the simple sentence, that is a sentence containing one subject and one verb. In the last example the subject is 'the Contractor' and the verb is 'obtain'. In order to make sure that the structure of the sentence is readily apparent, it is necessary to see how to analyse it. Analysis in the grammatical sense is likely to be cursory in this context, but it is only a form of an algebraic expression:

(Sentence) minus (subject) equals (predicate).

If the specifier thinks of it in that form, the separation of the important elements is easy.

Types of sentence

Before the structure of the sentence is explored further, it is as well to ask: why analyse? The answer has to be that analysis is necessary on this superficial grammatical level because the meaning of a complicated or obscure passage of text can really only be unravelled if the subject can be identified. Analysis also depends on the understanding of the other forms of the basic sentence structure.

There are three forms of sentence structure other than the simple form – double, complex and multiple sentences. The double sentence is constructed by joining two simple sentences with a *conjunction*:

The Contractor shall issue a Change Order Request *and* the Client shall acknowledge such a request within three days.

In this example there are two subjects ('Contractor', 'Client'), together with their two verbs ('issue', 'acknowledge'), complete with their predicates linked by the conjunction 'and'.

The complex sentence in the following example is more than just two sentences linked by a conjunction:

> The Contractor shall request a Change Order in writing if he considers that the work has been varied by the Client.

Here there are three verbs ('request', 'considers', 'varied') and three subjects ('Contractor', 'he', 'Client') with the word 'that' acting not as a conjunction but as a *relative pronoun*. A relative pronoun is simply a pronoun which is used to introduce or relate a describing (relative) clause to a subject. The complex sentence is bound to be more dependent on the relative pronoun than the double or multiple sentence; and because relative pronouns are not always correctly used, problems of understanding often arise in the complex sentence.

The multiple sentence is just the result of linking two or more simple sentences and one or more subordinate clauses with conjunctions, as in this example:

> The Contractor shall give the Engineer four days' notice of his intention to test the Plant, *and* the Engineer shall attend such test *and* inform the Contractor of his intention.

This sentence uses the *understood subject* in which the word 'Engineer' is not repeated in the last part. The specifier needs to be wary of this device, because it is vitally important in contracts that it is clear who is doing what. If in any doubt, the specifier should repeat the subject.

The object

Although the predicate must contain a verb, the verb does not have to have an *object*; thus the predicate does not need to have an object. Although such a statement may be trite, it is quite often that people look for an object in a complex sentence and fail to find one. For example:

> The Engineer/shall decide.

This has a predicate but no object, whereas the following has a predicate with an object:

> The Engineer/shall issue a *decision*.

In a simple sentence this may be easy to see, but in a complex sentence the problem may be more difficult:

> The average density obtained from groups of three determinations carried out in accordance with BS 1377 shall be not less than 95% of the

theoretical density of the material as compacted to zero air content calculated from the specific gravities, determined in accordance with BS 812 and the nominal proportions of the constituents, including the water.

The object itself comes in two forms: direct and indirect. The *direct object* appears in the following example in which the 'decision' is the direct object of the verb 'issue':

The Engineer shall issue a *decision*.

But, in the following, the direct object is absent, leaving only the *indirect object*:

The Contractor may appeal to *the Engineer.*

Or, as in this example, 'approval' is the direct object and 'from the Engineer' the indirect:

The Contractor shall await approval from the Engineer before proceeding.

Verbs

All sentences must have verbs. This simple rule seems to be lost on a number of specifiers, who are apt to write a line like:

Copies of Purchase Orders to Client on Fridays.

No doubt, they will argue that the line is clear. Indeed the line itself is quite clear, but the construction of a document on the basis of sentences lacking verbs would prove impossible. The only exception should be where the specification is intended to be a list of actions or things. This listing style of specification can be quite effective and appropriate, as here:

Copies of Purchase Orders to Client on Fridays
Expediting reports the next Tuesday
Inspection reports each Tuesday
Exception reports on Thursdays

Or:

Tiles, red concrete
Gutters, black plastic
Downpipes, black plastic

Although none of the lines forms a sentence, each is effective because the specifier is consciously writing a list. However, it becomes bad writing if the specifier slides into the following:

> Purchase Order copies must be sent to Client on Fridays. Expediting reports the next Tuesday. Inspection reports delivered each Tuesday. Exception reports on Thursdays.

So, if the specifier has decided that the verb does have a place in the specification, it is as well to know how to treat it.

Types of verb

There are four types of verb: the *finite verb*, which has to agree with its subject; the *non-finite verb,* which is a verb with no subject; the *transitive verb*, which takes a direct object; and the *intransitive verb*, which does not. Specifier are primarily interested in the finite and non-finite verb forms but should remember that the verb 'to expire' is intransitive, one can only 'expire' and one cannot 'expire (something)'. In the following paragraphs the words *participle, infinitive* and *gerund* are used before their meanings are explained in later paragraphs. The reader may wish to skip forward to these later paragraphs for descriptions of these terms.

Tenses

Finite verbs agree with their subjects, they are limited by tense and they conjugate, that is the endings of finite verbs change to suit the relationship of the verb to its subject. Therefore, the question of tense must be clearly approached. In building tenses, two forms of the participle are used – the present and the past. For example, with the regular verb 'to decide', 'deciding' forms the *present participle*, and 'decided' is the *past participle*. Participles have a number of functions: they act as adjectives, stand in absolute constructions and form tenses with the auxiliary 'to be'. The tenses of the verb 'to decide' are given fully in Table 6.2. The specifier will see that tenses can be thought of as existing throughout a time continuum – from the *perfect* through the *present* to the *conditional perfect*, the last tense representing the highest level of uncertainty. The time continuum can be expressed in two forms – the 'snapshot' action, such as 'I decided', and the continuous action, such as 'we shall be deciding'. However, the continuous tenses still relate only to discrete parts of the time continuum such as the past 'we were deciding'. In order to show the continuation of an action throughout the continuum, an adverb is needed:

> We have been working *continuously.*

Table 6.2 The tenses of the verb 'to decide'

(a) Active forms

	Past		Present		Future*		Conditional	Conditional perfect
	Past perfect	Past	Present perfect	Present	Future*	Future* perfect	Conditional	Conditional perfect
Simple form tenses	I had decided	I decided	I have decided	I decide	I/we shall decide You/they will decide	I/we shall have decided You/they will have decided	I/we should decide You/they would decide	I/we should have decided You/they would have decided
Progressive (continuous) tenses	I had been deciding	I was deciding	I have been deciding	I am deciding	I/we shall be deciding You/they will be deciding	I/we shall have been deciding You/they will have been deciding	I/we should be deciding You/they would be deciding	I/we should have been deciding You/they would have been deciding

(b) Passive forms

	Past perfect	Past	Present perfect	Present	Future*	Future* perfect	Conditional	Conditional perfect
Simple form tenses	It had been decided	It was decided	It has been decided	It is decided	It will/shall be decided	It will/shall have been decided	It would/should be decided	It would/should have been decided
Progressive (continuous) tenses	It has been decided	It was being decided	*It has been being decided*	It is being decided	*It will/shall be being decided*	*It will/shall have been being decided*	*It would/should be being decided*	*It would/should have been being decided*

Note:
See the text on p. 92 regarding the use of the passive form. The examples marked in italics are obscure to the point of being of little use – they are included in the table for completeness.
*See 'will' and 'shall' on p. 128.

The tenses that appear in Table 6.2 are all *indicative* tenses, but there exists an alternative form, rare in modern English, the *subjunctive*. The subjunctive is used only if the condition is unlikely or the event unlikely to occur. Although the specifier could argue that all contracts deal with events that are unlikely to occur, the subjunctive is as rare in contracts as elsewhere. The specifier usually states matters categorically and therefore in the indicative:

> If the Contractor fails to submit samples within ten working days, the Engineer shall...

The subjunctive form of the verb 'to fail' in the above example would be:

> If the Contractor were to fail...

As part of the time continuum the specifier will note in Table 6.2 the two *conditional* tenses, which are formed with 'would' or 'should'. There are other ways of expressing an action whose likelihood of occurrence is dependent on another action. There are degrees of certainty. For example:

- 'I may/might decide'
- 'I might have decided'

- 'I could/can decide'
- 'I could have decided'

An example of the form of the verb that utilizes the verb 'to be able to' as an auxiliary, as in 'I could/can decide'/'I could have decided', is:

> The Engineer could have approved the bituminous coating if proper test certificates from the manufacturer had been supplied.

He did not approve because of the *condition* that the proper test certificates were not supplied.

The infinitive

In common with the gerund and the present and past participles, the infinitive does not *conjugate*: it does not agree with a subject. The infinitive is simply the 'label' of the verb, that is the form that identifies the verb 'to decide', 'to allow', and so on. The verb cannot be labelled 'decide' because the form is not identified. There is obviously no verb 'is'; it is the verb 'to be'. The infinitive of course is that which cannot be split. The example known to all science fiction followers is:

> To boldly go...

The old grammarians would have leapt upon this and the following more specification-related example:

> To simply decide is easy enough.

Here the adverb 'simply' has come in the middle of the infinitive. Worse still is the construction:

> To simply (in the specifier's own mind) and irrevocably decide is easy enough.

The rule has its base in the attainment of clarity. The first construction could have been understood as 'Simply to decide...', implying that the act of decision itself is simple, or 'To decide simply...', implying that the object of the decision is simple. Carelessness in this matter could cause the specifier unnecessary trouble.

The infinitive form with 'to' has other functions. In the following example 'to rescue' forms part of the subject of the verb form 'is':

> *To rescue the programme* is impossible.

Alternatively, it can be the object, as below:

> The Client and the Contractor shall agree *to change the order of fabrication...*

Here 'to change' forms part of the object of the verb 'to agree'.

As an adjective the infinitive can qualify a noun, as in the following example:

> The Contractor shall have warehouses to store pipes, fittings...

Here the infinitive form is part of a purpose clause, meaning 'in order to store pipes, fittings...'. A further example is:

> The Contractor shall submit a Schedule of Welding Specifications [in order] to show their intended areas of application.

Participles as adjectives

In common with the infinitive and the gerund, the present and past participles do not conjugate, that is they do not agree with a subject, but they can function as adjectives. For example, in the following the present participle of 'to weld' is the adjective qualifying 'procedures':

> The welding procedures must be approved...

And the past participle qualifies 'joints' here:

> Welded joints shall be subject to…

The gerund

In common with the infinitive and the present and past participles, the gerund does not conjugate: it does not agree with a subject. It is a much misunderstood form of the verb. But the gerund simply is the present participle used as a noun, and when it is used in this way it functions like any other noun. The specifier will probably not distinguish between the gerund and the present participle. In this example 'welding' is the subject of the verb form 'subjected':

> The welding of nodes shall be subjected to…

Of course, as both the gerund and the present participle are (or can be in the case of the participle) nouns that can be qualified by adjectives, they demand that the verb agrees with them.

'Will' and 'shall'

Specifiers need to accept that the use of the future tense is crucial in the writing of sentences like the following:

> The Contractor *shall* test the Plant by 18 December and the Client *will* issue a test certificate within two weeks of the test.

This example, though simple, illustrates the correct grammatical use of the future tenses, which can be expressed in tabular form (Table 6.3). The use of 'will' and 'shall' in the future tenses interchange to show command, promise, or expression of determination. In the above example the contractor has a duty to test the Plant, but the client has expressed only the possibility of an action in the future, that of issuing a certificate. However, the specifier may have written:

Table 6.3 The correct use of' 'will' and 'shall' in the future tenses

Future	Command/determination
I shall	I will
You will	You shall
He will	He shall
We shall	We will
You will	You shall
They will	They shall

The Contractor *will* test the Plant by 18 December and the Client *shall* issue a test certificate within two weeks of the test.

Here is expressed the possibility of the contractor testing the Plant, but, if the contractor does so, the client is bound to issue a test certificate; no other condition needs to be satisfied apart from the test taking place. This is quite a different picture from the first example. Specifiers must therefore separate tense from intention, deciding whether they are to express a future state or make a future action compulsory.

'Must'

Surprisingly to some, the specifier can avoid the 'will/shall' debate entirely by using the verb form 'must', which does not conjugate or vary in any way: I must/you must/he must/we must/you must/they must. The verb form 'must' is clearly the expression of intention, making sentences such as the following quite unambiguous:

The Contractor *must* test the Plant by 18 December and the Client *will* issue a test certificate within two weeks of the test.

Phrases and clauses

The *phrase* is a group of words that does not form a sentence on its own, does not make complete sense on its own, and does not contain a verb. An example of a phrase is:

However, after the issue of the Certificate of Fitness...

In this chapter on grammar, 'clause' is explained in its strict grammatical sense. The word 'clause' is used elsewhere in this book in the context of a specification clause. A clause in a specification might consist of several sentences, a paragraph or possibly several paragraphs (sub-clauses). A *clause* (in grammar) has a verb and therefore a subject and predicate of its own. A clause may form a complete sentence if the sense allows. For example, the following clause makes complete grammatical sense:

The Client shall issue his approval of the welding procedures...

Such a clause is known as a *main clause*. The other type of clause is the *subordinate clause*, which, although having a subject and a predicate, does not make complete sense. The following, for example, does not make sense on its own:

... if the Contractor can show that the welding procedures were obtained on a previous contract with the Client.

However, the combination of the main and subordinate clauses makes a complete sentence, and complete sense:

> The Client shall issue his approval of the welding procedures if the Contractor can show that the welding procedures were obtained on a previous contract with the Client.

For the specifier, the subordinate clause is where the trouble usually starts. The relationship between the main and subordinate clauses is the key to producing a workable sentence. Of course, the specifier can avoid the whole problem by writing in simple sentences so short that he or she requires no subordinate clauses; but that is surely a defeat. The specifier must think of the relationship, choose the linking word between the clauses with care and ensure that the verbs and their subjects are not separated by such distances as to make the sentence incomprehensible.

Phrase and clause forms

Phrases and clauses come in the forms of the noun, adjective and adverb. As in the case of single words, a phrase or clause is what it does. Here a *noun phrase* provides the subject of the verb form 'is required':

> *Observance of the welding procedures* is required.

Similarly, this *adverbial phrase* qualifies the verb form 'produce':

> *Before the end of Stage I,* the Contractor shall produce a detailed programme.

And in the example below the *adjectival clause* qualifies the word 'certificates':

> Certificates *produced by the Contractor* shall be logged.

Subordinate clauses

The most complex writing comes in the use of the subordinate clause. Subordinate clauses relate the action in the main clause to some consequence in the subordinate itself. Linking words (e.g. 'who', 'that', 'because') are the key to the subordinate clause, but are insufficient to identify it because of the multiplicity of clause types, particularly adverbial clauses.

If specifiers remember that a clause is what it does, they will appreciate sentences in which the subordinate clause qualifies a word or group of words in the main clause or other subordinate. In a complex sentence this chain of clauses may be quite long, but the price of length is increasing complexity and decreasing comprehension. When tackling the construction of clauses or trying to analyse the meaning, attention must be focused on the words being described and the function of the word that links any two clauses.

Adjectival clauses

In the following examples – there are two clauses in each – the specifier can see the adjectival function in each case. In the first example the *adjectival clause* acts as an adjective describing the subject ('weld test pieces') of another clause:

Weld test pieces shall be prepared *which must include all the test specimens*...

Here it describes the object ('trial procedure test piece') of another clause:

The Contractor shall set up and weld a trial procedure test piece *that is in accordance with*...

In a more complex sentence the subject or object often becomes separated from its verb by the adjectival clause. Here the subject ('welder') has been separated from its verb ('permitted') by the adjectival clause:

A welder *who fails to meet these criteria* is not permitted to re-test until he has completed the period of retraining specified in...

Whereas in the following example the direct object ('list') has become separated from its verb ('submit') by an *adjectival clause* describing the indirect object ('to the Purchaser'):

The Contractor shall submit to the Purchaser, *who has right of rejection,* a list of consumables prior to testing.

The specifier may conclude that the linking words of an adjectival clause are: who, that, which, whose, whom, as, where, when, why. However, these words alone are insufficient here as it is the adjectival function that matters. In any case, the linking word 'relates' the subordinate clause to the main clause, hence the term *relative clause*. The linking word itself can be introduced by a *preposition:*

Welding procedures shall only be valid for a period of twelve months, *after* which they must be renewed.

Antecedents and the adjectival clause

The group of words that is being qualified by the adjectival clause is the *antecedent*, and it is the separation of the antecedent from the adjectival clause itself that can give rise to confusion of meaning. For example:

> Consumables shall be required to deposit weld metal having a maximum tensile strength of 650 N mm^{-2} and a minimum tensile upper yield and ultimate tensile strengths of the minimum guaranteed values specified for the base steel plate, which shall also exhibit adequate fracture toughness per section 8.6 of this specification.

Here the antecedent appears to be 'base steel plate', but it was intended to be 'consumables'. Specifiers can easily run into this separation problem either by allowing sentences to become too long (as in the example above) or by not placing the subordinate clause near enough to its antecedent in cases in which a long sentence is unavoidable. In the previous example the problem has two solutions, either:

> Consumables *which exhibit adequate fracture toughness per section 8.6 of this specification* shall be required to deposit weld metal having a maximum tensile strength of 650 N mm^{-2} and a minimum tensile upper yield and ultimate tensile strengths of the minimum guaranteed values specified for the steel base plate.

Or, alternatively:

> Consumables shall be required to deposit weld metal having a maximum tensile strength of 650 N mm^{-2} and a minimum tensile upper yield and ultimate tensile strengths of the minimum guaranteed values specified for the steel base plate. They shall also exhibit adequate fracture toughness per section 8.6 of this specification.

In the second case the pronoun 'they' is sufficient as the subject of the second sentence as 'base steel plate' is singular – but, if it had been plural, it would have been necessary to use 'consumables' to avoid confusion. This illustrates the need for specifiers to be clear in their own minds what they intend to modify when using the relative clause construction.

Adverbial clauses

The *adverbial clause* acts as an adverb, qualifying a verb or an adjective or another verb in another clause. An apparent complication in adverbial clauses is the fact that grammarians contend that there are so many types. However, the specifier may relax in the knowledge that the adverbial clause performs a function and its classification can be largely ignored. Nevertheless, the specifier may wish to know that the four main types are: time, place, manner and degree. No purpose is served by setting out examples of each type. More to the point is showing the construction itself:

The Contractor shall be responsible for all losses, shortages and damage *whilst the material is under his control.*

And:

The current revision of the drawing index shall be submitted to the Purchaser, *as the Purchaser requires,* but not more often than once a month.

Or, in a complex sentence in which the *adverbial clause* qualifies a previous subordinate clause ('specially lapping... paintwork'):

Painting repairs shall comprise the application of all coats to the original film thickness, specially lapping the coats progressively on to the flash-blasted adjacent sound paintwork, *so that this area is completely painted by the application of the final coat.*

A particular form of the adverbial clause, though not a type in itself, is this exception:

The Client will provide the pile grout system's specified equipment *except that the Contractor shall be responsible for including the grout injection tappings on each pile guide.*

The specifier can recognize the adverbial clause by its linking word: that, except, though, as if, where. But because some of these words also introduce adjectival clauses, the specifier should remember that *a clause is what it does.*

In terms of the reader's understanding, misplacing the adverbial clause is just as serious as misplacing the adjectival clause. For example:

A stick sample at least 20 mm diam. and 100 mm long shall be cast, labelled and retained by the Contractor for possible electro-mechanical testing *when the anodes are cast from each charge.*

It is not clear here whether the casting, labelling and retention of the stick sample or the testing takes place when the anodes are cast. The specifier could have overcome the problem by a rearrangement of the sentence to place the *adverbial clause* near its antecedent ('stick sample'):

When the anodes are cast from each charge, a stick sample at least

20 mm diam. and 100 mm long shall be cast, labelled and retained by the Contractor for possible electro-mechanical testing.

Noun clauses

A *noun clause* acts as a noun, and therefore it can take on the role of the subject, here telling 'what' satisfies the client:

That the welders had been recently tested would satisfy the Client.

Or can take on the role of the object, here answering the question 'what' shall the purchaser decide:

The Purchaser shall decide *when the Supplier may proceed*.

Linking words are important – what, that, when, if, that which – especially when the noun clause is the subject. When acting as the subject, the construction is commenced with the linking words, as here:

When the Contractor commences work shall be the 'Start Date'.

Noun clauses are also distinctive from the point of view of punctuation because they do not require separation from the main clause by a comma in any event. In fact this is quite logical as nouns themselves are not separated from their verbs by commas. Noun clauses also exhibit the implied 'that', as here:

The Client implied [that] *permission had been granted*.

In addition, the noun clause excites grammarians with some subtleties of construction that are rarely found in specifications.

Unfortunately, confusion can arise between a noun clause and an adjectival clause, not that the specifier really needs to be overconcerned about the distinction. In the following example the clause 'that uses a gas shield' is an adjectival clause describing the gerund 'welding', which itself is the object of the verb form 'permit':

The Purchaser will not permit welding *that uses a gas shield* in outdoor locations.

Whereas the clause is a noun clause acting as the direct object of the verb form 'permit' here:

The Purchaser will not permit *a gas shield being used* in welding out of doors.

Clauses and phrases as definitions

The punctuation of clauses and phrases is something that the specifier needs to be constantly aware of. Apart from the need to punctuate (especially with commas), the use of punctuation sets the meaning of the phrase or clause with respect to the rest of the passage. This paragraph sets out the punctuation for defining and non-defining phrases:

> Two samples *for chemical analysis* shall be taken from the pouring stream (top samples), *at the beginning and end of each charge*, except for charges of less than 1000 kg when one sample only need be taken.

The defining phrase is 'for chemical analysis' and it is thus not separated by commas from the noun ('samples') it qualifies. The non-defining phrase 'at the beginning and end of each charge' is placed between parenthetical commas. The specifier can think of the concept in terms of being able to remove the non-defining phrase without ruining the meaning of the word being qualified, the antecedent. Clearly, if 'for chemical analysis' is removed, the antecedent 'samples' loses most of its meaning.

Clauses also should follow the same rule, for example:

> The template jacking system employs three hydraulic jacks *which are mounted in the main structure supporting mud mats*, as indicated on the drawings.

The defining clause is 'which are mounted... mats', more fully describing the antecedent 'hydraulic jacks'. The non-defining clause, 'as indicated on the drawings', is (effectively) contained within parenthetical commas (although actually completing the sentence).

Punctuation generally

The use of punctuation is crucial to the meaning of any piece of text. Some of the model forms of conditions of contract discussed in Chapter 2 still limit the amount of punctuation in their clauses to the occasional full stop; others, for example the FIDIC form, use the full range of punctuation. The old-fashioned lawyer managed to avoid it entirely; often documents contained no punctuation from beginning to end. Whether this was born out of fear of getting it wrong or truly had another purpose is probably an unanswerable question. What is certain is that it is no longer the style of documents and therefore is not an option open to the specifier. The specifier needs punctuation, but there is no need to scatter it about the text like confetti. The old joke about the parliamentary draftsman who when asked what he had done that morning replied that he had inserted a comma, and when quizzed on his efforts in the afternoon replied that he had removed it, shows that punctuation can be supremely important at all levels.

The full stop

Full stops come at the end of sentences and in the middle and end of some types of abbreviation. There are no other uses of the full stop; they do not, for example, come in lists, as here:

Tubular and tubular transition sections
Tubular intersections
Structure general site assembly

A list is either part of a construction which itself requires punctuation or stands by itself, in which case it does not need any at all, even at the end. In the next example, in which the list is part of another construction, punctuation is needed:

This section of the Specification describes construction and testing procedures for

– tubular and tubular transition sections
– tubular intersections
– structure general site assembly
and related requirements.

Alternatively, the sentence might end in the list itself, as here:

The Contractor may claim if:

the programme is affected,
the manhours will be exceeded,
or the overhead allowance is diminished.

But when must the sentence end, requiring the full stop? On the basis that the sentence should contain only one idea and its development, the specifier should keep sentences as short as possible. Full stops should not be used in any headings or footnotes, though footnotes can be long enough to require complete sentences. There is usually no need to double the full stop when using inverted commas or parentheses. Finally, because there is no differentiation in typing or printing between the full stop and the decimal point, the use of decimal numbers in the text may be awkward. To avoid confusion, it is best not to end a sentence with a numeral.

The comma

As there are few rules on the position of the comma, its misuse can usually only be judged on the degree of interruption to the sense or flow of the text in question. However, there are a few rules, and two false rules that the specifier needs to identify:

- Never put a comma before 'and' *False*
- Do not separate adjectives with commas *False*

The question of the use of the comma before 'and' does not concern literature, but it could be crucial in the interpretation of a specification. In the following example the meaning is that the three actions, quite separately, are banned:

> The Contractor shall not A, B, and C.

However, if the specifier has omitted the comma, there are only two banned actions – A alone, and B and C together:

> The Contractor shall not A, B and C.

On the question of separating adjectives, the rule is that adjectives that qualify a noun and complement each other should not be separated by commas, but those of different senses should; the following examples illustrate the two forms:

> ... grey plastic gutter pipe

> ... heavy, metal object

In the latter example the job of replacing the linking conjunction in 'heavy and metal' has been done by the comma, whereas it was never necessary to write 'grey and plastic'.

The parenthetical use of commas is a common construction, being essential in making the distinction between defining and non-defining phrases and clauses. If specifiers use commas to bracket words, then they should remember what they are doing and not leave out the second comma. The specifier can use commas as parentheses, as here:

> The Contractor shall submit, prior to qualifying weld procedures, a Schedule of Weld Specifications and drawings.

A caution on the use of commas in this way needs to be added – problems can arise in the agreement of the verb with its subject. In the following example there are two possible forms of punctuation that require different verb endings. In the first the subject of the verb 'to need' is 'the Contractor', therefore it is singular in form:

> The Contractor, and Suppliers where applicable, needs to submit his monthly report by the second Friday of the month.

Whereas in the second there are two subjects, therefore the verb is plural:

> The Contractor and Suppliers, where applicable, need to submit their monthly report by the second Friday of the month.

The use of commas in the various types of clause appears under the relevant headings in the section on clauses above.

The semi-colon

The specifier can think of the semi-colon as a stronger version of the comma. It is useful in text to break up the staccato effect of full stops, and the semi-colon allows one sentence to deal with two or more related ideas, both equal in importance, for example:

> The Contractor shall provide reports on the progress of fabrication, identifying problems that may cause delays to the completion of the Works; photographs are acceptable for this purpose.

The semi-colon is also useful in lists in which there needs to be grouping within the list, and if commas were to be used, they might cloud the meaning:

> The Contractor's Quality Assurance Group shall be responsible for ensuring that the following are presented to the Client – drawings and specifications; certificates, release notes, and goods received notes and other documents required for materials; welding procedures; welder qualifications; and all aspects of inspection and testing.

The colon

The colon has a fairly restricted use in specification, though in other writing it can be used to good effect. The specifier commonly uses it in the introduction to a list, for example:

> A statement shall be submitted giving full details of:
>
> dayworks and materials,
> amount payable to date,
> amount invoiced to date,
> amount paid to date.

But it should be noted that there is no need to use the colon with a dash, they are two separate punctuation marks; the specifier should never double up on punctuation.

The colon can also be used to break a sentence in which the second half of the sentence expands or summarizes what came in the first half, for example:

> Only paint which is delivered in sealed containers, bearing the name of the manufacturer and properly labelled as to its type, batch number and instructions for use will be acceptable: paint shall not be used after the date of expiry marked on its container.

Because of the need for the specifier to avoid repetition, colons are not usually required in the explanatory mode.

The dash

The dash is very useful for marking a pause or break before a sudden change of direction in the sentence. The specifier may note that the dash can be used to introduce examples of all kinds in the form of inset text. In this form there sometimes exists the rare need to use double punctuation, as in the following case in which there may be some doubt whether the list of items appears distinct enough:

> Each shipment of anodes shall be accompanied by the necessary documentation to provide at least the following –
>
> – charge numbers included in shipment
> – numbers of anodes in each charge
> – chemical analyses for each charge
> – numbers and weights for those anodes weighed as per section 19.8 above.

However, the specifier can achieve the same effect by numbering each item, though that might affect the overall paragraph numbering.

The above example is one kind of change of direction, but the dash is useful even when no list follows. For example:

> The Contractor shall arrange spoil dumps on either side of the main access road – rainwater run-off must not be allowed to flow on to the neighbouring property.

The question mark

Specifiers should not resort to asking questions in their specifications, so all that needs to be said is that the question mark ends a sentence in the same way as a full stop.

Inverted commas

There are two sorts of inverted commas, unfortunately used for similar purposes. The specifier will need to use them consistently, but it is not easy.

Inverted commas are theoretically used in pairs, either single, '...', or double, "...". Inverted commas are used to mark out quotations and unusual words, including common words used in an unusual way. The specifier will rarely need to quote; nevertheless it is worth pointing out that the inverted commas (usually the double form) totally enclose the quotation, as here:

> Section 3.1.7 of the commissioning specification specifies that site welded joints in steel pipelines shall be tested "at a pressure of 24 bar for a duration of 15 seconds".

The other punctuation (in this case the full stop) comes after the closing inverted commas, unless it is part of the quotation itself.

Should the quotation run for more than one paragraph, the specifier must start each paragraph with the opening inverted commas, but only close when the quotation is complete; this is necessary to prevent confusion over the extent of the quotation, for example:

> The Contractor proposes that the following specification be taken from a previous contract –
>
> "Polyurethane enamel to give a minimum dry film thickness of 35 microns.
> "White anti-condensation paint to give a minimum dry film thickness of 35 microns.
> "White stove enamel to give a minimum dry film thickness of 35 microns".

In addition, the specifier will often use common words in an uncommon or special way, perhaps in some form of secondary definition as here:

> The Supplier shall deliver 'ex works'.

In all of the above specifiers are free to choose between double and single inverted commas. If they take to using double for quotations and single to show uncommon use, then provided that they are consistent no confusion should result.

Apostrophes

One use of the apostrophe is to indicate that letters have been omitted; for example 'it is' becomes 'it's'. Another use is in the *possessive*, which is more important to the specifier. There is a move afoot to lose the use of the apostrophe in the *genitive*, or possessive, case. This has been picked up by at least one 'quality' newspaper. Attractive though it may be in the newspaper world, newspaper editors do not have to suffer minute analysis of their words in cases of dispute. Consider these two examples:

> Contractor's (including subcontractor's) plant, to its full value, shall be insured by the Client.

> Contractors' (including subcontractors') plant, to its full value, shall be insured by the Client.

Now look at the alternative version:

> Contractors (including subcontractors) plant, to its full value, shall be insured by the Client.

The first two examples have completely different meanings: the first statement refers to a singular contractor and subcontractor, the second to multiples of both. The third example may mean either. By starting the sentence with the word 'contractor', the specifier cannot make it clear whether it is 'the contractor's' or 'contractors'' (similarly for the subcontractors) plant that is to be insured; without the apostrophe it is impossible to tell whether the reference is to a single contractor or many.

In the matter of the possessive the specifier must be fastidious in the use of the apostrophe – before the 's' in the singular noun, and after the 's' in the plural noun. In the case of words that already end in 's', or are themselves already in the plural form, the apostrophe comes between the two 's', though the specifier will not often need the phrase 'Dickens's books'.

Possessive adjectives – my, his, her, its, ours, theirs, yours – have no apostrophe, for example:

> The aggregate shall be checked for moisture content before batching commences and its moisture content shall be...

Parentheses

The common name for parentheses is 'brackets', but brackets come as the square or curly types. The ( ) are parentheses, and it is these that the specifier should use in text to contain an idea that is not part of the main thrust of the sentence. Specifiers should use square brackets only when quoting from another writer and they need to add a word of their own to make sense of wording out of context, as here:

> Total dry film thickness... [shall be] 290 microns minimum.

The curly type { } has its place only within mathematical formulae.

Specifiers should use parentheses to enclose one phrase or clause on the basis that, if they need to enclose more, then they should be writing more than an aside. Additionally, the specifier has to differentiate between the use of parentheses and parenthetical commas, which is difficult. A guideline to the latter is to restrict the use of parentheses to ideas that really are asides, and certainly avoid enclosing otherwise quite respectable phrases and clauses. Without doubt, the specifier should not enclose the defining phrase or clause (see p. 129) in parentheses.

If the parentheses themselves end at the end of a sentence, the full stop must come outside the last parenthesis, unless the enclosed text forms a complete sentence and has a full stop of its own (which is unlikely), for example:

> This information may be available as hard copy or in an agreed computer data form (disk or CD-ROM).

Punctuation should not appear within parentheses, and if it is necessary the specifier is probably trying to enclose too much text.

Capital letters

Though not strictly part of a discussion on punctuation, the correct use of capital letters is an area of importance to the specifier. A capital letter always follows a full stop, question mark or an exclamation mark: no other punctuation mark requires a capital, except inverted commas at the beginning of direct speech, for example:

> He said, 'The contractor is behind schedule.'

Reported speech does not require inverted commas, and therefore capitals are not needed, as here:

> He said that the contractor was behind schedule.

Proper nouns require a capital letter, for example 'France' or 'President'. But in the area of technical writing in which the specifier is involved there is a special class of proper noun, namely the defined term. In order to distinguish the defined term from its non-defined brother, the capital letter is necessary. What must be avoided is text looking like, perhaps, a piece of German (German nouns all have capitals), as in the following:

> The use of Flux Cored Arc Welding without Gas Shielding shall be subject to the Approval of the Purchaser.

The hyphen

The hyphen and the dash are quite different; the dash has a space either side of it but the hyphen does not. The most common use of the hyphen in specifications is in breaking words at the end of the line. If specifiers do this, they must make sure that there is at least a complete syllable on each line. The best guide for specifiers is to imagine reading the document aloud, then decide where a break would be sensible. For example, the following are reasonable breaks: 'con-struction', 'fitt-ings', 'there-after'; but these are not: 'wa-ter', 'ce-rtify', 'he-re'. Fortunately, the ability of word-processing software is such that lines can be proportionally justified without the need for hyphenation (see p. 156). The need to use a hyphen to break a word will occur only if the text is written in narrow columns. The specifier can also use the hyphen as a linking mark to join two or more words to form a single expression or word: 'high-alumina', 'sandy-clay', 'rise-and-fall'.

Another use of hyphens is in the separation of a prefix from a proper noun: 'pre-October', 'post-Award'. There is also the class of adjectival phrase that needs a hyphen to avoid misunderstanding, such as 'Contractor-supplied', 'Purchaser-approved'. Hyphens are also used in the separation of prefixes if the unhyphenated word would be ugly or unrecognizable, for example 're-enter', 're-weld', 'semi-independent'. However, the specifier should not split words that are already considered as single words, e.g. 'mismatch', 'equidistant', or 'preformed', and in cases of doubt should refer to the spelling and grammar tool on the word-processing software, or better still – a dictionary.

Abbreviations

There is no longer any need to insert full stops in abbreviations. The specifier can plead 'open punctuation' if challenged. For example, the specifier need not write 'A.S.T.M.E' or 'B.S.', but need only write 'ASTME' or 'BS'. However, full stops are still included in abbreviations of Latin words or phrases such as 'e.g.', 'etc.' and 'i.e.' (See also the section on 'Use of acronyms', p. 108.)

Writing formulae

Writing mathematical formulae has become much simpler now that modern word-processing software can cope with all the necessary typefaces, including symbols. Nonetheless, the specifier should be aware of a few guidelines as aids to clarity.

The specifier should set formulae in the middle of the line for short expressions. In the case of formulae that will stretch over more than one line, the line breaks should come just before a $+$, $-$, $=$, or $\div$ sign. The relevant sign should start the next line except where / (the solidus) is used for division.

Reference marks to other parts of the page (e.g. footnotes) which the specifier needs to put in the formulae should not be mathematical numbers or symbols; other symbols, such as *, §, †,‡, should be used.

Numbers

Numbers, or more strictly numerals, which are the symbols expressing numbers, come in two basic forms – Roman and Arabic. Arabic numerals are used almost worldwide, even in written Chinese and Russian texts, but oddly enough not in written Arabic. Roman numerals come in three forms, which explains their popularity in indexing:

1 *Full capitals:* I, II, III, IV, XX, XC.
2 *Small capitals:* I, II, III, XX, XC.
3 *Lower case:* i, ii, iii, xx, xc.

Arabic numerals are mathematically more versatile, but only have two forms – as numerals or as words (the word form of Roman numerals is not used).

Neatness in numerals should be of special interest to the specifier. If the specifier is careless in the use of numerals, it is apparent to the most casual reader, though it has to be said that it is probably of little contractual importance. But it might make life easier if there is a dispute over meaning; for example the specifier writes 'nine' but the contractor claims that it should have read 'twenty-nine'. If, in this case, the specifier can show through consistent use of numerals that if 'twenty-nine' had been meant then '29' would have been written, then the contractor's case is lost. This can introduce a set of rules, or rather a consistent way of being consistent:

- Use numerals for numbers of ten or less, except for zero in which case write 'zero' or 'none'.
- Use words when describing the number of times, e.g. 'fourth time' or 'four times'.
- Do not mix words and numerals in one amount.
- Use commas or spaces to separate numerals consistently, for example '1 456 345' or '1,654,432'.
- Write numerals in words if they introduce a sentence whatever the value.
- Use words for indefinite amounts, e.g. 'twenty-three or twenty-four days'.
- Use words for the time of day when not using 'a.m.' or 'p.m.' or 'hrs'.
- In dates or page numbering use the least number of numerals '1982–84' or 'page nos 81–7'.
- Always use numerals when writing decimals.

Fractions

The way that the specifier wrote fractions in the past was wholly dependent on the fact that typewriters offered only a limited range, commonly $\frac{1}{2}$, $\frac{1}{4}$, $\frac{3}{4}$.

Nowadays word-processing software can offer any range required. Therefore, specifiers have to be consistent in the application of their own rules. Some suggestions are:

- For common fractions of numbers less than one write as fractions, e.g. $^1/_2$, $^1/_4$, $^3/_4$.
- Alternatively, for common simple fractions write them out, e.g. 'two-thirds', 'three-quarters' (noting the use of the hyphen).
- For compound fractions use words, with a hyphen joining the last two words, e.g. 'seven thirty-seconds', or numerals, '$^{19}/_{600}$' or '$^6/_{32}$', but not '$^7/_{32}$nds'.
- For numbers greater than one, use words throughout, e.g. 'seven and three-sixteenths'.

Sentences and paragraphs

Specifiers can do a great deal to make their writing clearer apart from the use of correct grammar. If the specifier uses short sentences, short words and sensible paragraphs, the writing will become clearer immediately.

In the way that a sentence contains a single idea, a paragraph should contain a series of ideas with a common thread. Deciding when a common thread of ideas is exhausted is surprisingly difficult; if the paragraphs are too short, this gives a very broken effect to the writing in a stylistic sense and also increases the danger of separating an action from its intended effect or consequence.

For example, in the following extract it is not clear to which 'approval' the 'approval *in writing*' in the final line refers:

The Contractor shall seek approval from Suppliers for the Contractor to pass on the Supplier's guarantee.

The Contractor shall obtain approval from all Suppliers, manufacturers, assemblers and fabricators for the Client or his delegate to visit any premises from where the Plant is to be delivered.

The Contractor shall obtain approval *in writing*.

To overcome that particular problem, the specifier should have included 'in writing' in each paragraph, even at the risk of repetition. Certainly, the specifier should introduce a new paragraph at each change of place, subject, or time. From the point of view of layout, a further discussion of paragraph length appears in Chapter 7.

7 Model specifications and layout

The White Rabbit put on his spectacles. 'Where shall I begin, please your Majesty?' he asked. 'Begin at the beginning,' the King said gravely, 'and go on until you come to the end: then stop.'

Introduction

Model specifications, in the guise of standard specifications (see Chapter 3), can be external to a company or they can be internal to a company. The 'Civil Engineering Specification for the Water Industry' is an example of a reasonably long-standing external form of model specification. However, in the context of this chapter, model specifications are particularly looked at as a tool for the production of particular (project-specific) specifications. The chapter also describes the standard frameworks and layouts for the production of specifications.

The power and sophistication of modern computers and of modern computer software is ever increasing, and specifiers now have very sophisticated tools at their disposal. There are many different types of computer software available to specifiers. Fortunately, commercial forces have led to a certain amount of standardization of software. The consequence of this is that the specifier is likely to be working with a Windows-based word-processing package of known or familiar capabilities. This chapter is not intended as a tutorial on any particular software package. It does look at how modern computer systems can be used efficiently in handling text and in the production of coherent specifications.

The seemingly unlimited power of modern computing tools can be a blessing in disguise as far as the production of specifications is concerned. It is important that specifiers become involved in the creation and efficient use of model texts and that these model texts are used in a simple and controlled manner. Thus, the specifier needs to remember that a computing system is now more than an electronic typewriter and corrector of text. In fact, the very property of easy correction is part of its failure to attract intelligent use. Seen in its proper role as a processor of text, in which processing includes library storage, it allows the specifier a new freedom to produce text efficiently and accurately. Properly used, the library facilities allow the specifier to be confident of his or her sources.

Once the specifier has grasped the idea of automatic assembly, this chapter provides guidance on the assembly of the drafts and access to the final document by contents pages and indexes.

Document management

Computer-based document management systems are widely used by organizations for the storage and control of their key documents. They are an alternative to paper files and are another step in that search for the holy grail of office work – the paperless office. Document management systems are useful tools for the efficient use of model specifications. The computer-based system is often an adjunct to, or it may fully represent, the company's quality, health and safety, and environmental management system(s). And, as such, it will be written and structured around the necessary procedures for that company's business. Model specifications represent a key resource for use on a company's projects, and they will inevitably have a place in any document management system. The specifier will be involved in deciding how the model specifications are stored, maintained, controlled and used within these systems.

Filing and finding

Whether the library of model specifications is a sophisticated document management system or a simpler database of computer files, the following principles for filing and finding model specifications hold good. Within computer software pieces of text of whatever length are usually referred to as 'documents'. Each document is allocated a unique reference within the system by the specifier, and it is important that the form of this reference is understandable and logical and not driven by any constraints of the computer-based system or by any constraints stemming from the wider quality systems of the company. This point is important because it may prevent specifiers from using the document label that is used within the computer-based system within any paper-based filing system. However, it is best to try to make the machine accept an external referencing system rather than to allow the referencing to be driven by the writer of the software system, who may know little or nothing of how to organize a retrieval system on the lines of a conventional library.

When specifiers create a document, it must be consigned to some part of the system in such a way that it is readily retrievable. Whatever the constraints with regard to the naming of files within the system, specifiers need to set up a library that has logic, like any filing system. If they do not, the storage capacity will be used inefficiently without any clear idea of how large the eventual storage will be, and this results in expensive and untidy storage.

Most organizations store their information on a central server with appropriate backup facilities, and total storage capacity is not such an issue as it was even 5 years ago. However, the existence of unlimited storage is no excuse for the library of model specifications to be structured in a sloppy and inefficient way. Therefore, the major constraint on the specifier is not likely to be running out of storage, but in the integration of the model specifications into the in-house management systems. In the case of a large client company, with in-house engineering capability, this can be quite difficult. The difficulties are fewer for a company whose main business is engineering or consulting.

To achieve efficient use of the central library facility on the system, specifiers must start with a set of rules that enables them to monitor growth in the library, and to set target dates for the revision of stored documents with some automatic 'flagging' device. The review periods can be set individually on certain types of document, for instance a minor works specification could be deleted as soon as the contract is awarded; the hard copy will be the evidence that is required for the future running of the job, and that type of specification has no value as a master. On the other hand, a major specification might need to be kept in case it can be used as a whole in the future. Both depend on the fact that the constituent parts are kept as model clauses that have been assembled, and perhaps altered, for the particular task in hand.

In order that users can access the library, a form of index needs to be set up, in plain English, to which an enquirer can refer to inspect the models on offer. It is not a good idea to make this index out of the document codes themselves, probably mixtures of alphabetical and numeric characters, but it should be made of keywords that the enquirer can see to be relevant. The machine can do the necessary cross-referencing between the keyword and the document code. For example, an enquirer who wishes to inspect the available specifications on road gravel is much more likely to recognize '*gravel, roads*' than '*etnI45.856*'.

Keyword indexes are a valuable tool in searching for the right specification because the computer can be asked to search for 'gra' instead of the full word 'gravel' and is therefore not faced with matching as many letters in each entry with the one that is required. If users are given a keyword index and the rules under which it has been drawn up, they will be able to carry out more efficient searches than otherwise. In all the setting up of the library, the specifier must make it as friendly to the enquirer as possible; otherwise it will not be used at all.

Model specifications

Within the central library of the system the specifier will set up the model specifications as individual documents. In most cases the models will only be single clauses ready for assembly. In some cases there may be a need for grouping of clauses which would be treated as a whole in any assembly, for example there may be complete preambles suitable for assembly into schedules of rates, or collections of general clauses suitable for the specification of concrete. However, the specifier is not aiming to store complete, standard specifications in the central library. The larger the groupings become, the more difficult it is to maintain the library and prevent the wholesale alteration of model text; the whole library then tends to become merely a store of 'what we did last time' and the integrity of the model text theory collapses. There is, after all, no such thing as a complete standard specification for a particular type of project; each is different. This is the essential difference between manufacturing industry (repeatability) and construction and site installation. As a house is built from standard bricks yet is different from all other houses, the specifier can assemble the specification from the models according to the need of the individual project.

The time taken to assemble a specification from choice each time is minimal, and not really an argument for maintaining complete specifications. That is not to say that, for example, a specification chapter on electrical wiring should not always contain particular clauses in a particular order, but the content of those clauses is based on the model and not the last job.

If there are to be true model clauses that specifiers can use with confidence, whether or not they originated them, they must have been created following known and well-considered rules. When specifiers are creating the models they must consider at least the following matters, and record them in such a way that they can be shown to users at any time:

- Who was the sponsor?
- Who approved the draft?
- Who authorized the issue?
- What is the review date?
- What special rules govern the use?

To avoid a massive buildup of paper in the system library, the specifier must include that information with each model in such a way that will not interfere with the assembly of the document as a whole. Most computer software will allow comments to be stored that will not appear in print, but that is really not much help because the user will want to see the authorities when he or she sees the draft. The facility for 'call out' boxes is available in many software packages and is a useful way in which to record such comments. Otherwise the specifier will probably have to resort to some trick to allow the comments to be printed, possibly by putting them at the end of each clause and deleting them in the final version (this can be laborious). The specifier can even take this to the extent that two versions of any specification are always produced – one with the authorities and comments, and one without. If this can be done easily, it may be worthwhile expanding each model to include reasoning and even instructions to the site staff as to how the clause is to be interpreted and why it was included in its present form. Of course, the version with the comments would not appear in the contract, but it can be of help on remote sites where it is not clear why something was specified in the way that it was. After all, the specifier has to manage and allow for changes to the specification, so it is better that those who are likely to change the specification know why it was drafted in the way it was before they start changing it.

In some organizations it will not be necessary to have two versions of specifications because the remote sites will have ready, and reliable, access to the central library by remote local area networks (LAN) or by the Internet. If such access is available, the optimum solution is probably to have a printed copy of the contract specification on site and to make available any commentary in electronic form. An example of the 'two-version' specification appears in Table 7.1.

Related specifications

The specifier must also deal with related specifications when creating each model because some clauses depend on the inclusion of relevant information from elsewhere, for example:

Sulphate Resistant Cement

[text]

If this clause is specified for works below ground in drainage applications specify also Bituminous Paint Coat

The specifier has also to deal with the cross-referenced clauses in a similar way. Naturally, the model cannot contain cross-references by number because there are no numbers in the models in the library. Most computer software will allow some sort of prompt to remind the user that a piece of text is missing, e.g. '... ?cross-ref: valve type? ...', or actually ask questions on the text that the specifier has to answer on assembly: '... ?Engineer or Employer? ...'. (The use of the double question mark simply serves to mark the beginning and end of the prompt from the rest of the text.)

Too many standards?

When the specifier is creating a model specification, especially when making references to standards, care must be taken not to make the model the sum of all known standards or to incorporate too many specialist views. Both will probably make the model contradictory. Another temptation is to make the specific into the general by the extrapolation of particular rules or standards into areas in which they are not applicable.

Table 7.1 A 'two-version' specification for pressure pipelines

Butterfly valves

Butterfly valves shall comply with BS 5155; certain clauses of BS 5155 are amplified as follows:

Clause 4.1, Type:	Double flanged
Clause 4.2, Service application:	Valves shall be suitable for tight shut-off application unless a regulating duty is required
Clause 13.1, Bearings and seals:	Shaft bushings shall be provided
Clause 15	Valves shall be mounted with shafts horizontal unless otherwise specified or shown on the drawings

INSTRUCTIONS TO SPECIFIER

Valves may be included in the specification for pipelines, but where there is a chapter covering Plant (e.g. in a pumphouse) consideration should be given to incorporating valves with those items.

NOTES TO SITE STAFF

Clause 4.1	Type alternatives are: (1) double flanged; and (2) wafer (three designs); the latter is a cheaper but less satisfactory option and was not chosen as low cost and small space was not of great importance
Clause 13	Bushes were chosen for interchangeability
Clause 15	Horizontal mountings suffer less from particle ingress

Initiating a model specification

There should be no restriction on who may initiate a model specification, and the specifier should provide any likely author with guidance on how to seek the necessary approvals. The intention should be to allow work to proceed from an uncertain start to an approved and certain end, though not all draft models that are started will reach the library. The specifier should not allow drafts to remain in that state for longer than necessary, they should either be approved or withdrawn.

Regular updating

In order that specifiers can maintain confidence in the use of the library that has been created, they must have a procedure that others can inspect and use if necessary; if the library is allowed to get out of control, then its value is diminished. In addition, if specifiers are to allow only authorized changes to the models in the library, they must set out instructions as to how those changes are to be made. This will involve security in the computer system itself and a backup system of approval by signature which will relate the change to any given authority. Needless to say, rapid issue of changed models must follow the changes themselves if the library is also held on hard copy in the office. But it is far better to maintain the computer-based system as the controlled version and any hard copies as uncontrolled versions.

Model format

All the models that are stored in the library must have the same layout or format in order that they present a consistent whole when the specifier assembles the models in a single document. The effort of reformatting text that has been rapidly assembled partly negates the speed of assembly. Specifiers must decide the layout that will apply to all their output and convince others that standardization is necessary. The preferred layout should then be stored as format instructions in the library.

Dated and undated standards

If specifiers need to make reference to standards, they will either make reference to a dated version or omit the date. Should the specifier make an undated reference, it means that the latest version of the standard is to apply on the basis that the organization issuing the standard is acceptable to the specifier, and the latest is acceptable in all cases. However, if this is the case, the specifier should state as much in a preamble, and also specify when the 'latest' issue cannot be allowed to supersede the last – usually at the point the design is frozen, or the tender is issued.

If the specifier quotes a dated standard, it means that no other will do, and no explanation of that is needed. Any waiver of that provision will have to be dealt with in the same way as any other change in the specification.

House style

To the user of documents, especially those who have never been involved in the production of them, it would seem that their appearance is just a matter of chance. However, the specifier will quickly learn (if he or she does not know already) that layout is no accident.

Specifiers will probably be constrained by 'house style' when they produce a document, but, even so, they should be able to obtain the style of document that they consider to be the most appropriate. They will also be constrained by the type of model form of conditions of contract which has been chosen for the contract. Some, particularly the 'design and construct' forms, include quite lengthy schedules in their layout. In the same way that the words that are used are planned, so should the layout be planned. The aim is to make the document meaningful for the user – and this is where the house style constraint often fails. If specifiers find themselves up against a house style that they consider to be badly conceived, then they should seek to change it; though this is often easier said than done.

Natural order of events

In any piece of text the specifier must have a plan for dividing the text into 'natural' sections. The use of the idea of natural divisions is important from the user's point of view. Layout in all sections of text, however small, should follow some sort of order that the user will recognize as natural – usually on the basis of importance or the passage of time.

Labelling the divisions

As there are layers of divisions of text, there is the problem of what to call them – i.e. divisions and subdivisions, and so on. There are many ways to label the main divisions – chapter, section, appendix, volume, and the more uncommon ones such as text and protocol. Only the word 'section' allows the naming of subdivisions in a related way. In the matter of subdivisions the specifier has further choice – subsection, paragraph, section, article, clause and item. Going further into the realms of the subdivision becomes increasingly illogical, although this does not apply to the same extent to the numbering (see p. 157). It is simply not worth the effort to name subdivisions (or rather sub-subdivisions) as 'the sub-sub-clause'; it is always best to stop the name at the first level, i.e. at 'sub-'.

Part of the planning in what to call the divisions of the work is how big to make the divisions themselves. The only real guide is the perceived attitude of the user. If the divisions are too long, then reference to the text will be cumbersome and involve searching through large blocks of type.

Page layout

A key element in the appearance of documents is the layout of the page itself. In the modern word-processing packages there are guides to the layout of text, but the specifier has to remember that the reason such guides are written is to sell computer software. Unfortunately, specifiers do not seize opportunities to influence the layout.

The first item to tackle is the size of the margins – the framing of the page – in this case, A4 size. The left margin exists for the purpose of binding, and is usually about 30 mm. The right margin also has this function if the document is to be bound back-to-back, and therefore should be of the same dimensions. The specifier must remember that any marginal headings, numbers, or notes must not fall into the binding area, otherwise the text will be forced into a narrower field. The top margin is usually shallower than the bottom one; there seems to be no particular reason for this, except that it prevents the text from appearing to slip off the bottom of the page. Accordingly, the top margin should be about 25 mm and the bottom about 40 mm. This spread of margins leaves enough room for fifty-five lines of text, with the page number in the bottom margin. The number of lines obtainable depends on the size of type used.

If the specifier sets the margins as above, there will be enough width for seventy-five characters on each line, depending on the type size. However, the number of characters per line obviously depends on the specifier's policy of indenting text from the margin, which itself depends on the numbering system that is used (see p. 157).

It is usual to leave line spaces at the end of paragraphs and under headings. In order to tie the heading to the piece of text that follows it, it is normal to leave a three-line space after the end of a paragraph followed by the heading of the paragraph, then a two-line space before the start of the text. As for line spacing within the text, the convention is that drafts should always be in double spacing (allowing room for corrections), with the final text in single spacing. However, the final spacing is very much a matter of personal convenience, usually depending on the specifier's view of the final size of the document. The more open appearance of the text afforded by a one-and-a-half spacing is often worth the extra paper.

Automatic page numbering

A function that is available on all computer software is the automatic numbering of pages in any part of the assembled document. It is not necessary to number from the beginning to the end, the specifier can nominate any start point for the numbering, and set the numbering to run with a title; for example:

Technical Specification 12-23

This means the 'twenty-third page of Section 12'. It could also be set as follows:

Section 12 Page 23 Technical Specification

As long as the title is less than a single line, there are few restrictions.

Automatic numbering is very useful, but the specifier should try to run series of pages in as short a length as possible, which usually means a chapter at a time. This avoids having to reprint masses of text if editing is required after the production of the hard copy of the document and the editing has caused repagination.

Pagination

The automatic breaking of pages into the required length is known as pagination. The software usually carries this out without reference to the person who is editing or typing, but there are times when the specifier needs to know or set the rules by which the software will carry out pagination.

In the initial typing process of any text that will carry over more than one page the machine will allocate to each page the number of lines that the specifier has set, remembering that lines include blank lines. Therefore, the machine is not concerned with the logic of breaking a page at any particular point. This leads to page breaks between headings and their accompanying paragraphs, and after a single line following the paragraph heading – known as 'widow' lines. In order to prevent this, the specifier can command that the carryover of lines shall never be fewer than two, and that there will always be at least two lines after a paragraph heading at the bottom of the page. Even this might not be acceptable, especially when paragraphs themselves are short, in which case there may be occasions on which only one line is carried to the top of the following page – known as 'orphans'. To overcome this, the specifier can instruct the software to 'block' the paragraph headings with a given number of following lines, say five, and, if this cannot be achieved, to carry the whole block over to the top of the next page. If the specifier uses this too strictly, pages of grossly uneven length will result, which may be as undesirable as widow or orphan lines.

One solution to unsatisfactory carryover of lines is to force a page break in the editing process, but this is tedious; it may, however, be of use when the specifier is constructing lists that might lose their effect if split in an arbitrary manner. During the editing process, especially if the specifier has forced page breaks in the text, the insertion or deletion of text will cause the pages to be paginated in a different way. Therefore, the specifier needs to reduce the length of page runs to a manageable size, so that the repagination comes to a natural halt, perhaps at a chapter or section ending.

Lists

Within the text, whatever the line spacing the specifier has chosen, there will arise sections that appear to need special treatment – usually lists of items and tables. In the treatment of lists the specifier needs to be consistent, and a useful rule to follow is to insert half-line spaces above and below lists. If listed items cover more than one line, the turnover lines can be indented in respect of the first line, as shown here:

> 12 m columns with 3Nr Type K
>   luminaires
> 12 m columns with 2Nr Type K
>   luminaires
> Auto/manual timeswitch
> Lighting distribution system
> Operating manual

However, in this matter it is not easy to demonstrate a satisfactory rule.

Tables

Simple tables are an alternative to lists and they can be a useful and clear way of presenting information in a specification. The formatting options available for tables are almost endless and excessive formatting of the boxes in the table is probably best avoided. It is far better to keep the formatting as simple as possible so that it does not overpower and distract from the content of the table.

Typeface

Of course, one of the most important items is the size and style of the typeface. In this matter there is almost infinite choice. The choice of typeface is now subject to interpretation by researchers and psychoanalysts, but this fact should not influence the specifier in his or her choice of typeface. The normal choices of Times New Roman, Arial or Book Antiqua are all appropriate for specifications. House styles will normally dictate a particular typeface within client and consultant organizations.

Justification of text

Justification of text is the automatic adjustment of the lines of text so that they neatly fit the space between the left and right margins. Alignment of text is different. Text can be aligned to the left or right margin or about the vertical centre line of the page. Aligned text has a ragged edge appearance along the unaligned side(s); for example, a right-aligned block of text will have a ragged edge appearance on the left side of the page. There are two ways that text can be justified – either by stretching out the words and spaces in a series of lines to produce a smooth right margin or by proportional spacing. The former method produces text with sometimes large, and always irregular, spaces between words, and allocating the same width to all letters, so that an 'i' is given as much space as an 'm'; it is really a pointless exercise. However, the latter method of proportional spacing allocates spaces throughout the text according to the width of letters, and the spaces between the words is varied to meet the requirements of the width of the line. This page is set by proportional spacing.

When the specifier chooses right-justification for the text, only those lines that are not stopped short of the end of the line by a full stop or other punctuation will be justified to the right margin; those lines that fall in lists and are inset (and do not reach the right margin) will have unjustified edges.

Headings

There is a need to plan the hierarchy of headings throughout the text in order to present to the user a logical sequence of division. The specifier can do this by using the available print faces, but not using so many that the reader becomes confused. The specifier can consider the full selection to be:

- **ALL CAPITALS BOLD**
- <u>**ALL CAPITALS UNDERLINED BOLD**</u>
- ALL CAPITALS
- <u>ALL CAPITALS UNDERLINED</u>
- **Lower case bold**
- <u>**Lower case underlined bold**</u>
- Lower case
- <u>Lower case underlined</u>

In addition, there are the various options such as italics and the expanded typeface. However, specifiers should not suppose that all the above options are available in a single document because (numbering hierarchy aside), if the specifier uses too many levels, users will not be able to tell at a glance what the hierarchy is.

Headings need not always be aligned to the left margin if the document is to be produced back-to-back. Aligning the heading to the right margin on the right-hand page is useful, but it means that the specifier has to set the heading positions at the final proof stage because any alteration during the editing process could throw the page from one side to the other. If it goes wrong, the result is a mess.

Paragraph numbering

In almost any document the specifier will have to use a referencing system for divisions of the text. How far the system is taken will depend on the length of the document, but mostly on the need for the user to locate a relevant part quickly. In most specifications there is a need for quite a fine level of division. But there is a price to pay in that the referencing system may become too complicated, even to the point of taking over the layout entirely. The following example is taken from a large project specification:

2.2.3.1.3   Line Pipe Material
2.2.3.1.3.1 Steel pipe shall be in accordance with API 5L...

There is little chance of relating this clause and sub-clause to their relative positions in the structure of the specification sections because of the complexity of the paragraph numbering.

Whether the specifier chooses a numbering system that is wholly numeric or alpha-numeric, or even wholly alphabetic, is not important (although the latter is very cumbersome). The important thing is how detailed the referencing system is going to be. Two examples are:

1. Section Heading

    1.1 Subsection

        1.1.1 Sub-subsection

A. Section Heading

    A.1 Subsection

        A.1.1 Sub-subsection

The above illustrates the danger of subdividing the sections using both the numbering system and the layout of the text. The habit of indenting the text is both wasteful of space and unnecessary. If the text uses indents as well as numbering, this is, in fact, double-referencing and the result is only wasted paper. It also makes impossible the option of using a fourth level of referencing because the third reference has pushed the text across the page to such an extent that the fourth subdivision will be just a column down the right-hand side of the page. The problem of the text creeping across the page becomes more apparent when the third or fourth level of division is a major piece of text and continues over several pages. Therefore, where there is a need to reference the text, the specifier should always bring the text to the left margin at all levels, as here:

1. Section Heading

1.1   Subsection

1.1.1   Sub-subsection

The specifier should continue to keep the paragraph reference to the left of the page, even when there is a list which does not form part of the referencing system, thus:

3.9   Requisitions to Purchase

Requisitions to purchase shall have attached the following papers:

(a) the request for prices,
(b) ˙the offer from the supplier,
(c) the responsible engineer's approval.

The use of bold for the paragraph number and name, or just for the number is a useful marker.

Automatic paragraph numbering

The paragraph numbers in the library of model texts will not fit exactly into the structure of every contract specification and it would be impossible to achieve this. Paragraph numbering can be indicated in the model texts and it can be fixed at the sub-clause level, but inevitably each section of the specification will require reformatting at clause level. The most reliable way to automatically set the paragraph numbering is to reformat a complete specification section using the numbering formatting style tool on the word-processing software. This involves a certain amount of input from the specifier, but it is a relatively foolproof system.

Line numbering

Line numbers can be added to the side of the page in the hard copy of the specification. The reason put forward for doing this is that they provide a useful aid for reference and location of a piece of text, particularly when the document is produced in single line spacing. However, line numbers are rarely, if ever, added to specifications, probably because they make the document look untidy. The addition of line numbers in the hard copy has the effect of cluttering the page and it detracts from the overall appearance of the document.

The compromise, if one is needed, is to make the document available in both paper and electronic forms; in practice, electronic copies of the specification are often provided. The word-processed version of the document provides information at the bottom of the computer screen not only on line numbers but also on the position along each line (column) and on the page number. On the electronic version of the document each character can be referenced uniquely.

Marginal notes

Marginal notes used to be quite common, particularly in model forms of conditions of contract. They are used as a means of explanation to the parties of the document and often rely heavily on cross-references to other clauses, but in conditions of contract they are always excluded from the interpretation of the clauses to which they are attached. This is a difficult concept for many people and it would be difficult to argue that marginal notes in a contract specification were not intended for use in interpretation of the specification.

Marginal notes would be an affectation in a specification and, not least, they would be difficult to produce and maintain. The aim of the specifier is to write a specification that is clear and self-explanatory. If further explanation is needed for the client's staff on site, then it can be provided in comments (see Table 7.1).

Sentence layout

This section is concerned not with the grammar of the sentence, but with its appearance. Apart from the length of sentences, which depends on a mixture of grammar and style, the specifier can only change the appearance of sentences by splitting them up on the page itself.

When using the 'schedule' sentence layout, a specifier should be able to construct complicated sentences (in terms of meaning, not grammar) more easily. The following example illustrates the point, taking the normal layout first:

> The Contractor shall prepare reports that will enable the Engineer to review monthly the work to be done against the time limits together with the work to be done in the given review period (normally one month), and that work which was intended to have been done but which remains uncompleted, plus the required remedial action, and the work remaining with corrective actions to be taken.

This is a fairly formidable piece of text, even when it is complete. But if the specifier changes the layout to that of the schedule sentence, the text immediately becomes easier to read:

> The Contractor shall prepare reports that will enable the Engineer to review monthly the following:
>
> (i)   work to be done against the time limits;
> (ii)   work done in the given review period (normally one month);
> (iii) work which was intended to have been done but which remains uncompleted and the required remedial action;
> (iv)   work remaining and the corrective action to be taken.

What cannot be illustrated is that the schedule sentence soon exposes when the specifier is rambling in a complicated explanation, or has embarked upon a potentially disastrous set of relative clauses that will, sooner or later, lose their antecedents. Nonetheless, the specifier is still constructing sentences and therefore must take care with the punctuation throughout the expansion down the page.

Document libraries and standard layouts

This section is not so much concerned with writing the text, but with handling text that has already been written. If the specifier is involved in repetitive work in the sense that a lot of the sections of work are common, then some sort of library base is needed. Hopefully, this base will not just be the last document that is produced, but a properly thought-out library. In such a library the importance of standard layouts cannot be overstressed. For the assembly of model texts into a document to be a real time-saver, the specifier must be spared the need to reformat texts of differing layouts. And it does not matter whether the original model text comes from a computer or not. Although computer assembly systems can assemble text very quickly, all the benefit is lost if the formats are incompatible – it is a tedious exercise to alter all the format commands. For example, one text may be set up as follows:

* left margin 10, right 5;
* letters per line 60, lines per page 60;
* headings lower/upper case bold;
* tabs at 5 10 15;
* two-level paragraph numbering.

And another set of text might be set up as follows:

* left margin 15, right 10;
* letters per line 65, lines per page 55;
* headings lower/upper case underlined;
* tabs at 6 12 18;
* three-level paragraph numbering.

Should the specifier try to merge these texts into a single document, especially with many samples from each, this will create work. The resulting text will not look neat unless all the margins and other parameters are the same. There is no point in setting up and using a library of model texts only to incur a reformatting exercise. The software tools are very powerful and it is not a major problem to reformat a document as a consistent whole. However, it still takes time, and efficiencies can certainly be made if the library of model specifications is entered to a standard format style The library owner must agree the style with potential specifiers and publish it – specifiers then have the responsibility to adopt the style.

Building the document

Once specifiers have satisfied themselves that a document can be built without any technical hitches, the planning and execution of the compilation can be confronted. The specifier must flesh out a body from a skeleton. In some ways the planning of the compilation has to be more thorough than with pure writing because there is little scope for crossing out and rewriting. Therefore, the first act is to become familiar with what the library has on offer by reviewing the keyword index and the short descriptions supplied with the library. The specifier should do this before inspecting any of the model texts and should then construct a plan, listing all the texts that are to be used and noting where the library cannot supply the necessary items.

Once the specifier has identified from the library the texts to be used, the compilation process is moved through quickly without inspecting the results too closely. This is important – if specifiers start to edit before all the texts are in front of them, this not only slows down the compilation process but will probably be working against the wishes of the original writers and missing the benefits built into the model texts in the first place. Once the specifier has all the texts ready for inspection, the editing process begins.

The next stage is to read through the complete compilation to check whether any suspected omissions still need drafting and writing. It is important, at this stage, not to be in too much of a hurry to cut out or change items which the specifier has retrieved from the library. Once specifiers have tackled the omissions, they can go through the other parts of the text. In the process the specifier should leave intact any notes to the user (see Table 7.1) or other references or authorities. Other commentators on the document will need to see that the specifier has assembled the texts and obeyed the instructions to the compiler. At this point, the first draft should be complete.

Model texts and prompts

A prompt in the model text is an indication that the specifier needs to take some action when using the model to make it into a full specification. A prompt is not a warning which relates to the specifier's use of the model, but an instruction to take some action: for example, the words 'do not use in Kenya' are a warning, not a prompt.

A prompt can take a number of forms, for example:

(needs matching inspection clause)...
(check the cross-reference)
(?valve type?)...

The specifier satisfies the prompt by writing in the correct clause, number or term.

In many word-processing packages the specifier can allow the machine to seek out the prompts for checking. If the specifier leaves the prompt code in the machine as long as possible, it should be possible to check the validity of the insertion in later drafts in which the text may have been altered by others.

Glossaries

Glossaries, as far as word-processing packages are concerned, are stores of standard texts – sometimes also known as auto text. The standard text can be inserted into specification clauses, which may, in turn, have come from model clauses, to reflect some particular feature of the contract. The most obvious example would be the contractual name given to the client – 'Employer', 'Buyer', 'Purchaser' or 'Client'. The specifier could play safe and always refer to the 'Client' in the specification clauses and include a defined term:

> Client means Purchaser or Employer or any other defined term meant to indicate the promoter of the Works.

Alternatively, the specifier can use a glossary term to tailor the specification clauses to match the phraseology used in the conditions of contract. Another example of the use of a glossary term in a model specification concerns the use of proprietary names for suppliers' equipment. A particular client might prefer waterproofing systems for concrete floors from a particular manufacturer, but another client might have no particular preference and the specifier can amend the relevant specification clauses appropriately using glossary terms.

The alternative to glossaries as an editing system is to use manually the editing tools on the word-processing software to replace chosen words with new words. However, this system is not foolproof because it relies on character recognition and can sometimes change parts of words which the specifier had no intention of changing. If several changes are to be made in this way, then the sequence in which the changes are done will also be important if any of the words are common. Glossaries are a more reliable way of automatically changing small sections of standard text.

Drawings

If the specifier is using model texts with associated drawings, perhaps themselves produced in a computer-based drawing system, the problem of the compatibility of text that appears on the drawings is not diminished; it may actually increase because of the ease of putting text on the drawings. It is important to recognize this point because site staff will use drawings in everyday use before they refer to the specification.

Authority for issue

The last stage in the compilation is the authority to issue the document, and the specifier would be wise to have kept all the internal references in the text up to this point, in order to answer queries. The procedure for issue of the specification as part of a tender enquiry document will form part of an organization's quality procedures and there will be clear rules governing the authority for issue. The use of model specification clauses by the specifier is compatible with such a quality procedure and it will go a long way towards simplifying the procedure.

The final act is to reference the document uniquely and insert the date. The latter point is, for some inexplicable reason, not usual with specifications. But it is particularly important if, for example, the specification refers to: 'The latest edition of the following British Standards:...'

Contents page

The contents page fronts the document. It is literally a list of what the document contains, often with page numbers; it is not, and should not, be referred to as an index (see p. 165). The contents page should be in sufficient detail to enable the user to see precisely what is in the document and where to find it. If the contents page does not give page numbers, section numbers may be used instead. The specifier must decide on the degree of detail to show, and between merely listing the chapters and listing every section within chapters there are a number of choices. If contents pages themselves become too numerous, there is always a danger that the specifier will not be able to maintain accuracy when the document undergoes successive revisions. If it is decided to go down to the fine details, then the specifier should consider putting a simple contents page at the beginning of the document, showing the chapters, and at the beginning of each chapter displaying detailed contents which show page numbers for each labelled section. In this way, alterations can be made within any section without altering the main contents page.

On the question of layout of the contents page in which chapter titles are short and the page is the usual A4 size, aligning page numbers down the right-hand side will result in a wide gap from left to right. The specifier can overcome this by setting page numbers close to the chapter title, or setting page numbers down the left-hand side before the chapter title. Leader dots going across the page from chapter title to page number are untidy.

Index

An index is an alphabetical list of items in the text, together with the page, paragraph or section references. It is usually found at the end of a document, although in multivolume documents it is more useful at the beginning. The majority of specifications do not include indexes. If the specification is well laid out and the contents pages are at the appropriate level of detail, it is quite simple for the majority of users to find what they are looking for in the specification. There is a danger in using an index that all the appropriate clauses concerning a particular topic might not have been included in the index. However, the index can provide an objective view of the document, but it is difficult for specifiers to be wholly objective if they have done most of the editing in the first place. The services of a professional indexer are unlikely to be always available. The index should always be preceded by a note describing the system the specifier has used in its compilation and the extent of the coverage of the subjects.

Basically there are two types of index: those using repetition of words and phrases appearing in the text and those using keywords as descriptive references. The latter is often made up of the section headings themselves. The title page, contents, summaries and similar pages are not included in the index.

An index can be quite complex, in that phrases can be cross-referenced within it. A simple form might look like this:

Ceiling boards 230
Ceiling linings 256
Cement mortar 197, 214, 234
Cement wash 199
Centring 201–11

A more complex form might be as follows:

Dry ground excavation,
 ground level by bulldozer, 118
 ground level by hand, 103, 106, 109
 ground level by mechanical excavator, 108–12
 shafts and basements, 123–5
 trenches by hand, 129–32
 trenches by plant, 126–9

The specifier should make the page references as precise and concise as possible, as in the above example. To write '3ff' does not help much. Multivolume documents should show the volume number as well as the pages, for example: 'Roads, II, 17–32'. Punctuation is not important as long as the specifier uses it consistently. Cross-references may be necessary to avoid too much duplication, but the specifier should try to avoid cross-references, for, in a way, duplication of the page reference (only in indexes!) makes the index easier to use. Where necessary, the specifier should enter cross-references thus:

Tie beams *see* steel beams

Building an index is laborious work, and they do not appear to occur much in specifications. However, much of the labour associated with the old card-index method of construction simply does not exist any more. Therefore, to build an index the specifier only has to go through the document and mark the words to appear in the index. The index can then be generated using the tools available within the word-processing package. The sort program places them into alphabetical order. The specifier edits the index by amalgamating repeated references to the same item. Take, for example:

Steel tubes 30
Steel tubes 89
Steel tubes 143

In the final index, the entry for steel tubes would appear simply as:

Steel tubes 30, 89, 143

Proper names are not usually included in the index, unless they are trade names:

Joints, watertight,
    concrete, in 22–30
    epoxy materials 23
    Serviseal 28
    water-bar 27

Definite and indefinite articles do not appear in indexes, but prepositions are more difficult. If their omission does not impair the sense of the entry, then the specifier should either omit them or invert the words, as shown in the above example. The following example shows equivalent entries for entries in which prepositions are not needed:

Without	*With*
Works	Works
access	access to
definition	definition of
fencing	fencing around

But here the preposition is essential:

Without	*With*
Works,	Works,
care,	care of,

As for arrangement, the specifier should choose the simple letter-by-letter alphabetic format because computer software will probably sort on that basis, for example:

Nickel
Nitre
Nitric acid
Nitro-cellulose
Nitrogen
Nitroglycerine

The alternative is a word-by-word index, which is more complicated to compile and has no particular advantage except in groups of related entries. The rules for the word-by-word index are that entries of more than one word are arranged by alphabetical order of the first word, those with the same first word are arranged alphabetically by the second word, those with the same first two words are arranged alphabetically by the third word, and so on. Hyphens are treated as spaces, producing separate words, for example:

Word-by-word	*Letter-by-letter*
Cement content	Cementation
Cement grout	Cement content
Cement render	Cement grout
Cement wash	Cementing
Cement–water ratio	Cementitious
Cementation	Cement render
Cementing	Cement wash
Cementitious	Cement–water ratio

For entries involving numbers or chemical symbols the job is altogether more complicated. The specifier should refer to BS ISO 999, *Guidelines for the content, organization and presentation of indexes.*

Correcting

Checking should go on at all stages in the preparation of the document and not just at the end. To make the process easier, the specifier should date all drafts to make sure that everyone is checking the same version and should not use methods of correction that obliterate the original. The process can be quite easily controlled if the specification is compiled within a document management system. Gone are the days when the specifier used correction fluid to cancel written errors or cut up pages and made replacements using a photocopier. Not all specifiers are aficionados of word-processing software, and most probably only use 10–20% of the editing and formatting facilities that are available in many packages, but in most cases this is more than enough to compile a well-formatted specification that is coherent and consistent with the house style.

Proofreading

Proofreading is more than just checking, it is a science, and few amateurs do it well. The specifier cannot assume that what he or she has written, and may also have typed, will be free from errors, and it is essential to ensure that the typed version is exactly as intended.

Basically there are two ways to proofread – alone or in pairs. The latter method cannot be justified on the grounds of cost. If it is believed that the task cannot be done alone, then it should be passed to a professional proofreader. Specifiers will find it difficult to proofread their own work, because they will know what is written and will naturally read at a faster rate.

The problems in proofreading alone are, first, the difficulty of maintaining concentration and, second, the tendency to read into the text the words that one expects to see. The matter of concentration is largely one of setting the environment straight from the start, with no telephone calls or other interruptions. In addition, the specifier should attempt to work in only short bursts of about an hour at a time. An amazing amount of proofreading can be achieved in an hour. Another element in maintaining freshness in the task is to tackle the text in a random order. There is no need to go through the text from beginning to end because one is no longer interested in the sense of the document, only in the accuracy of presentation.

When proofreading from a hard copy, proofread a phrase at a time, with the original draft on the left (if the specifier is right-handed) and the text to be corrected on the right. Always keep the phrases to be read short, and ensure that the original text is read first. It is quite easy to end up proofreading the wrong way round. If necessary, a ruler can be used to keep to the place on the page, but of course this then entails proofreading line by line which is not as efficient (or easy) as phrase by phrase.

Proofreading marks

If the specifier has used somebody else to type the whole document (and this is rarely the case in a modern office) then time can be saved by using standard proofreading marks. However well the specifier and a particular typist may understand one another, the whole operation will flounder if there is not a standard method of denoting corrections that is applicable to all concerned. The learning of standard proofreading marks is now one of those dying skills that is worthy of resurrection.

There are two standard proofreading conventions, which are covered by the various parts of BS 5261, *Copy preparation and proof correction*. Part 1 of the Standard deals with the design and layout of documents and Part 2 covers the specification for typographic requirements, including marks for copy preparation and proof correction. Part 3 deals with the same issues for mathematical copy. The disadvantage of adopting the Standard is that everyone must be familiar with all the symbols, and neither specifiers nor typists are usually involved with printers to the extent that such learning is easy or worthwhile. It seems to be common to use the former Standard's reliance on words and symbols together. The specifier will probably wish to continue this practice, and so specimen texts appear as Figures 7.1 and 7.2, using the marks of BS 5261 and marginal words where explanations are commonly given. Only the more common marks are shown.

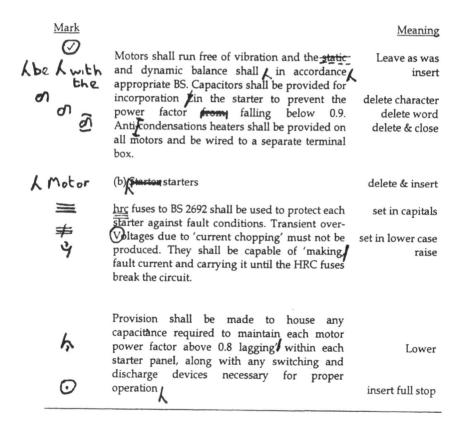

Mark   Meaning

Motors shall run free of vibration and the static and dynamic balance shall in accordance appropriate BS. Capacitors shall be provided for incorporation in the starter to prevent the power factor from falling below 0.9. Anti condensations heaters shall be provided on all motors and be wired to a separate terminal box.

Leave as was
insert

delete character
delete word
delete & close

(b) Starter starters

delete & insert

hrc fuses to BS 2692 shall be used to protect each starter against fault conditions. Transient over-voltages due to 'current chopping' must not be produced. They shall be capable of 'making fault current and carrying it until the HRC fuses break the circuit.

set in capitals
set in lower case
raise

Provision shall be made to house any capacitance required to maintain each motor power factor above 0.8 lagging within each starter panel, along with any switching and discharge devices necessary for proper operation

Lower

insert full stop

Figure 7.1 Proof-correction marks for deletion, insertion and substitution of text

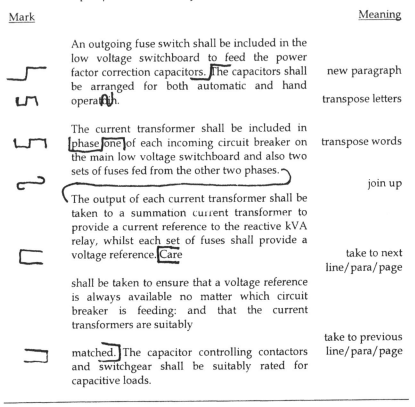

Figure 7.2 Proof-correction marks for positioning and spacing of text

Checking cross-references

Cross-references are best avoided (see p. 101). If they are used (and they nearly always are) they will need to be checked before issue of the document. Many word-processing software packages provide for the automatic insertion of cross-references, but the final document should be checked for correct cross-referencing before it is issued. The specifier should treat the checking of cross-references as a separate exercise from proofreading in order not to disturb the flow of reading. A good method is to mark off with a highlight pen the cross-references as each one is reached at the proofreading stage and then to check them afterwards.

Proofreading from a VDU

The specifier will probably find it easier to work from a piece of printed paper, but VDUs have improved greatly and many specifiers are quite happy to proofread documents from a screen. This has the advantage that some of the software tools can be used to highlight errors. Useful tools on word-processing software cover:

- spelling;
- grammar;
- tracking changes;
- comparing different versions of a document.

However, the specifier should be aware of the software settings and not be afraid to exercise the right to over-ride the suggestions from the software.

Choice of bindings

Basically there are three types of binding available: the reopening, the lie-flat and the close bound with either plastic or stitching. Of the three, the most inconvenient to use is the close bound because it prevents proper perusal of the document in an open position. As specifications are intended for regular use as contract reference documents, there is no point in binding the specification in such a way that will discourage use. If the specification is not expected to need updating by insertion, the lie-flat ring binder is quite satisfactory. Alternatively, the plastic ring-bound spine provides a functional and smart document, particularly when used in conjunction with a professionally prepared front cover, complete with company logo(s). This form of binding comes in sizes suitable for a wide range of document thickness. However, documents bound with a plastic ring spine which are greater than 18 mm thick can end up looking very tired and dog-eared towards the end of a lengthy contract, and should a 30 mm document fall apart at some stage, it can be a thankless task to reassemble it.

If specifiers intend that amendments be inserted into the document over a period of time, then they should choose a binder that can be opened easily such as the metal-ring 'D' binder. This will also make it easier for contractors to copy the document, or parts of the document, for issue to subcontractors and others who may not have ready access to e-mail. A plastic ring-bound document is notoriously difficult to copy as the numerous punched holes often foul the photocopier.

8 Claims

'You couldn't have it if you did want it,' the Queen said. 'The rule is, jam
tomorrow and jam yesterday – but never jam today.'

Introduction

There are several books on claims and much literature in the form of magazine
articles and conference notes – adding to all this may seem like gilding the
lily. However, little of the writing on claims tackles the problem at root –
the clarity of the original requirements of the specification. There is also the
point that all claims of a purely technical nature must be tested against the
whole set of contract documents, even if the primary cause of the claim is to
be found in the specification.

This chapter is intended to provide guidance on how specifiers should
react in situations in which their specifications are queried. In addition, the
specifier will discover that much advice elsewhere avoids the issues both of
the client's reaction to claims and the client's pursuit of claims against the
contractor; this chapter, hopefully, will go some way to redress the balance.

When the specification has survived the drafting, redrafting and approvals,
and the critical attention of tenderers, it will come into operation. No
construction project can run without necessary questions being put by the
parties involved concerning the content of documents that are in use on the
project. The specifier needs to recognize that some of the problems arising
out of the specification are worthy of attention in order to save the client
money. Contractors have an acute awareness of profitability and how
ambiguities in specifications can affect this, but all too often clients react
defensively to maintain the specification in the last resort. Contractors will
react positively to clients who pursue their interests in a correct and logical
manner.

Partnering

There is clear evidence that the increase in partnering in contracts has reduced
the number and frequency of claims. The very fact that parties are willing to

enter into a working relationship which is overtly one of cooperation must lead to a reduction in the number of claims.

The claim – what is it?

In normal everyday discussions on construction projects the word 'claim' is bandied about loosely to describe any form of query from a contractor – or it is suppressed in order to discourage the contractor from making any. The logic of the latter point is obscure, just as the inexactness of the former is pointless.

In a fair contract the allocation of the consequences arising from things not going according to the original agreement is based on a very simple principle, that is the person or body at fault will carry the liability so that the other party is not out of pocket because of that fault. If the cause of the breach is neither party's fault, and this is usually captured in the *force majeure* clause, then each party bears its own costs and the contract periods are adjusted accordingly. In this way the contractor is not unreasonably subject to liquidated damages for delay, and similarly the client would gain relief on any time-related milestones, such as the release of part of the site, if these were affected by the event of *force majeure*. All projects are subject to changes – and change gives rise to formal notification by one party to the other by variations orders, change orders, scope changes, and so on. Most model forms of contract provide mechanisms by which certain variations can be made in the normal course of that contract. The client, the Engineer or the Project Manager would usually make such changes or variations. Occasionally, the contractor might have powers under the contract to recommend variations. These are changes that both parties agreed to allow when they entered into the contract, and if the mechanisms in the contract work as intended and all parties accept that this is the case, no claim is involved. Claims occur when change occurs outside these contractual provisions. By the formal notification of a change, the other party is asked to agree to the change being incorporated in the contract. Throughout this process no claim is involved; it is merely the exchange of request and agreement, offer and acceptance. The bargain undergoes modification with the consent of the parties. If, however, the client and the contractor do not agree on how to modify the bargain, or on how the change will be implemented, then they will register their lack of agreement. This is done by one party submitting a claim.

But not all claims arise out of changes. The contract may not require *any* changes at all, but one party may think that the other is not adhering to the contract as it is written. The contractor may not be complying with the specification, or the client may not be paying what the contractor thinks is due. If the obligations under the terms of the contract by one party to the other are not met, the aggrieved party will seek compensation – i.e. requesting more time, more money, reperformance, or any combination of all three.

Such requests are claims and may be put forward by either party: the contractor or the client.

The problem with claims starts with definition. No doubt all specifiers will recognize this statement but may argue that the company insists on all queries on payment or time being registered as claims and handled accordingly, or that nothing is to be registered as a claim until it is clear that the parties cannot reach agreement. But in order to suggest how the specifier should deal with claims it is necessary, for the purpose of illustration, to maintain the widest possible definition, independent of 'house rules', of their exact classification.

A specifier's role

A specifier is involved in claims – either by formulating them or by analysing them. The assumption in all that follows, however, is that the specifier does not take the initiative in originating or deciding claims, and that both these functions are taken on by the specifier's management or the Engineer (if the contract so provides). The specifier is required to bring to bear his or her knowledge of the technical intricacies on the origination or resolution of the claim. The reason for the specifier's limited scope in this is simply that all claims, however technical, are based upon the contract as agreed and therefore are ultimately contractual in nature.

Supporting the contract as written

On the premise that all engineering contracts are written contracts, within the documents will be found the necessary words to support any particular line of reasoning on any matter arising during the construction. Of course, such a statement is patent nonsense – not only because documents are imperfect, but because the common law gives protection and imposes liabilities alongside the actual terms of the documents. However, from the specifier's narrower viewpoint the analysis of the claim (whether to support its submission or to decide on its acceptance) has to be based on the provisions of the documents and the ensuing exchanges of instructions, drawings, etc. arising out of them.

If the specifier has to originate a claim without a contractual basis, its acceptance will have to be by a separate agreement or an ex gratia settlement, the latter being a payment or dispensation of some kind to which there is no strict entitlement. There can be no philosophical objection to subsequently agreeing that a state of affairs was not foreseen at the time of entering into the contract and thereafter arriving at a settlement. But if a claim is based on the documents, the specifier can only reject it on the documents – there is no possibility of rejecting claims out of hand, the rule being only that the party who asserts must prove. In some cases a claim drags on because the party putting it forward refuses to admit at an early stage where their claim originates – inside or outside the written terms of the contract. To attempt to squeeze claims into the contract is not a good use of intellect.

This all leads to the point that the specifier, whether or not actually involved in the drafting, must support the contract as written, and that includes recommending changes if what is written does not meet the circumstances. If the specifier reads into the specification that which is not there, either as a defence to a claim or as preferential engineering, this will cause problems. Specifiers cannot consider what they (or others) meant to write; if they do (from the client's point of view) the client will incur greater costs than would have been incurred if they had accepted the apparent mistake in the first place and varied the contract appropriately.

From the particular to the general

All claims arising out of the specification will start from a particular problem – for example a cracked tank, a bowed wall, or a process which fails to operate properly. That is not to say that there are not other forms of claim of a much more general nature – for example disruption by other contractors, delay by one party to another, problems which are quite unforeseen and beyond either party's control *(force majeure),* and so on. But specifiers are assumed to be concerned that the specification is adhered to in each particular, and it is from the particular that they must start.

If specifiers are faced with a failure in the contract, they should go to that part of the specification that deals with the failure. A failure claim cannot be originated on the basis of 'shoddy workmanship' or 'general inability' – there must be particular failures to catalogue. However, from the catalogue of failures may come certain general conclusions, but even those are not 'general' in such a way that a general liability under the contract flows from them without involving legal argument – but that is not the specifier's aim.

In studying the specification in connection with a failure on the part of the contractor, specifiers should ask themselves five questions:

1 What is the exact nature of the failure?
2 What do the words actually say?
3 Was the specification changed during construction?
4 How (if at all) were approvals given?
5 What was the result of the failure?

From this list the specifier can assemble the data necessary for the acceptance or rejection of the claim. Notice that none of the questions addresses the financial aspects of the claim; that is an exercise which the specifier will probably delegate, but in any case, it is done when the form of the claim itself is decided. In fact the claim may be accepted or rejected before any monetary calculations are done, even to the point of a decision at arbitration or in the courts. The validity of a claim does not depend on its value – though that may be a spur to pursue or defend it appropriately.

Tactics: answering a claim

What comes as the biggest surprise to specifiers when they are faced with a claim is to be told that there is no need to answer it. There is no need to reply to a claim unless the contract specifically sets out a claims procedure which (of course) must be followed. The specifier should always acknowledge a claim and do any such registration of it that the company requires. But the fundamental point of any claim is that it must have a sound factual and contractual base, and that is what the specifier must establish before he or she replies to it. The specifier should not be rushed into answering (and presumably acceding to) a claim just because the other party presses it strongly. In the real world the strength of personalities sometimes outweighs reasoned argument based on fact. In any case, to be rushed into rejecting a claim can be just as bad but for a different reason – it may give the other party the idea that all claims will be rejected out of hand.

None of this should be taken to mean that delaying the resolution of claims is a good thing – it most certainly is not – but it is meant to draw the specifier's attention to the need to allow enough time to investigate the claim and sometimes, quite incidentally, to treat a number of claims as linked events.

Linked claims

If a party to a contract is alert to incidents that affect its performance of, or benefits from, a contract, the party is likely to make a claim as soon as an incident occurs, without waiting to see whether any other events flow from the first. Many claims are linked and can be dealt with together. If the specifier encourages resolution of a claim too early, a later claim from either party may nullify the conclusion of the earlier, and a complicated chain of set-offs may result. This is not to say that the specifier should use this reasoning as an excuse for unnecessarily delaying the resolution of a claim. If it is clear that all the events that might be connected with a cause have arisen, the specifier should encourage a speedy resolution of the claim.

Registering a claim

The essence of a claim is to make the original move while the evidence is still fresh – either in physical form, or in the form of instruction or cause of delay. In many contracts there is a time bar to making a claim in order to ensure that the claim is presented before all recollection of events is lost.

When registering or notifying a claim, specifiers should only be concerned that they have the makings of a valid claim by checking the specification thoroughly to see under what provisions the claim will be pursued. Clearly, if specifiers cannot find any provision under which to put a claim, they have a problem to which there are two possible solutions – either abandon the claim, or move from the particular to the general. The latter course needs careful thought, for it will probably mean a move out of the specification and its exact requirements to either the payment sections of the contract or the conditions of contract. In either case, expert assistance will be required.

Once specifiers have identified the particular provision or provisions, they should draft support for the claim in the most general way while maintaining the reference to the clauses, for example:

> The buoys deployed by the Contractor were lost between April and June and as a result no current data was obtained from the west sector. In accordance with Specification Sections 1.6 and 1.8 performance of that portion of the work on the west sector is incomplete, and payments should be reduced accordingly.

This is sufficient to notify the other party that a claim for compensation is to be presented. Alternatively, if there is an element of reperformance, the specifier might lay it out like this:

> The gearboxes attached to MPS pumps Nos 3, 4, and 6 exhibit vibration in the range 600 to 700 rpm. Specification Sections 1.7 to 1.10 set out the criteria for vibration-free operation up to the working speed of 1500 rpm. The Contractor's contention that since there is no vibration at the working speed the specification has been met does not over-ride the apparent failure to meet the test specification at all run-up and operating speeds (Section 1.8.5 refers). The Contractor should be instructed to provide new gearboxes which meet the specification.

The specifier needs to note that in both of the above examples the references are only to those parts of the specification that have apparently not been met, and they are in the form of an internal memorandum to the person who will draft the covering letter to the other party. The separation of the two functions is beneficial because it allows the specifier to concentrate on the technical aspects. If he or she does so, the subsequent memorandum can then be read against the conditions of contract or payment terms, possibly allowing the actual claim to be widened, for example to deduct liquidated damages or claim against a performance bond. In any case, the covering letter to the other party can refer to an attached memorandum.

Allowing work to proceed

The process of making or answering claims is not an end in itself, but rather should be a parallel activity to the real work of the contract. Because of the many external pressures that affect construction projects, there is always an element of risk in balancing getting everything perfect and settling for less provided that it does not negate the overall aim of the contract.

In cases in which the client may wish to reject unsatisfactory work but the contractor contends that either the work is in accordance with the contract or adjustments to design will accommodate a local failure in construction, there may be a true difference of opinion which cannot be resolved within the programme constraints. In these cases the specifier is required to do two things: first, to carry out the normal analysis of the failure to meet the specification, and second, to check the contractor's solution. If the pressures are such that work will proceed whatever the outcome of the analysis, the specifier should recommend that the decision to proceed is at the contractor's risk and that the specifier (if he or she so believes) will not recommend a waiver of the specification. If the advice is acted on by the client, then at least the client will not have thrown away any rights under the contract.

However, it may be the case that the contractor will not proceed to programme unless an assurance is obtained from the client that the remedial work, if done to an agreed specification, supersedes the original specification, and thus the contractor will not face a claim that he did not adhere to the (original) specification. In these cases the specifier's task is no different from that in the previous paragraph – the analysis still must be provided – it is only the client's negotiating position that has changed.

Programme pressures on the specifier have to be recognized as quite separate from technical pressures, and it is best to bring them together at the management level rather than delegate them to the specifier at the technical level.

Specification and modifications

The specification, however much the specifier who did the original drafting may regret it, is not a static thing – it will have to respond to the demands of the construction process by undergoing changes. In all claims the effect of the changes that may have taken place will have to be considered in the analysis.

When faced with a claim based on the original specification and 'site variations' the specifier must check that the site variations were properly authorized under the terms of the contract. If they were not, the specification remains as it was, and the contractor has carried out work either outside the terms of the contract or, perhaps, forbidden by it. For the very reason of avoiding any argument over whether changes were or were not authorized, the specifier will have ensured that a simple procedure for change notices will have been incorporated in the general part of the specification. As a result, the analysis of the modified specification is made easier.

If it is the case that the contractor should have proposed modifications to the specification, as is often the case in 'design and build' contracts, but did not, the client's specifier will have to research the claim from a different angle. The client in such a case will want to base the claim on the failure of the contractor to modify correctly or originate a correct specification as well as adhere to it during building. That is a more difficult task than pursuing a contractor for non-adherence because the client's specifier will probably be in the position of having delegated the detailed specification to the contractor. Thus, the specifier may be, say, in the position of testing a process which does not work, and then having to decide whether it has failed because of design or construction faults. A joint effort with the construction supervision team will be needed to check whether the specification was followed during building. As in all cases, the precise contractual liabilities only flow from the facts surrounding the failure and the provisions of the contract.

Effects of correspondence

Apart from the question of whether the specification was modified only in accordance with the contract, is the point of how much the parties to the contract allowed a state of affairs to develop, in seemingly perfect agreement, to the point of failure, at which one party stands on its contractual rights. This can happen in the frenetic activity of construction projects and there may be no clear point at which the specification was breached. This produces a real problem of analysis of any claim arising out of such situations. In the end, it may be that both parties will be deemed to have waived their contractual rights and will have to live with the situation as it finally unfolds. Nonetheless, specifiers should concern themselves with listing authorized modifications to produce the true modified specification and then follow the 'unauthorized' changes to their end-result. By doing this, specifiers enable other advisors to decide more easily where liability for the resulting failure lies. Such a situation is almost bound to end as an argument based on a study of the more general liabilities in the contract.

Warranty schedule

In a project in which there are many suppliers and contractors there is a need to keep track of all the guarantees and warranties in an orderly fashion. In a contract with a single contractor there will usually be one 'guarantee' – correct completion of the contract – although the contractor may in turn have received guarantees in purchase orders on his suppliers.

When commissioning takes place in stages, it is usually very difficult to persuade suppliers to give any guarantee that depends on a date of commissioning rather than date of delivery, let alone an open-ended guarantee. Naturally, suppliers take the view that they have no control over the commissioning date. Thus, if the client has to store plant for so long that the guarantee expires, it is hardly the supplier's fault. On the other hand, such guarantees are not much good to the client. There are ways round the problem such as maintenance guarantees or even insurance. However, the need for records is obvious, and the warranty schedule is a convenient form.

The warranty schedule is a means of overcoming a potentially difficult problem. If the commissioning period is lengthy, some items of equipment may have been in use for 6 months before overall commencement of the defects liability period. The warranty schedule will need to contain the following information:

- order number;
- equipment identification number;
- equipment title;
- description;
- vendor's name and address;
- date delivered;
- date commissioning due;
- date commissioning achieved/attempted/failed;
- defects liability period (from/to);
- date of any defect repairs;
- revised defects liability period (from/to);
- special conditions of warranty.

In the event that there is an equipment failure, the specifier can quickly obtain all the necessary information within which to frame the technical assessment and recommendations.

Insurance claims

The specifier will become involved in insurance claims when the 'contract object' has been damaged in some way. The 'contract object' might be a building, a piece of equipment or a whole plant, but it is unlikely to be the product of pure design – that is, drawings or paper. However, it is possible that computer data storage might come into the category of 'contract object' in the context of the insurance claim. As far as the specifier is concerned, the insurance cover is likely to be 'physical loss or damage' cover rather than the professional indemnity type which covers design failures.

Small insurance claims on a large site can be quite numerous, depending of course on what is the minimum allowable claim – the majority of contractors carry a substantial excess on their insurance policies. But in all cases of loss or damage there needs to be a consistent way of recording the occurrence. A 'loss and damage' register is a method for consistent recording, for example:

- Incident number
- Location, date, time
- Brief description of incident
- Equipment affected
  – plant identification number

- supplier
- purchase order number
- Probable cause
- Weather conditions (if outside);
- Damage to third-party property;
- Injury to persons
  - names
  - employers
  - hospitalization details
- Test pieces and analyses (laboratory name and report numbers)
- Photographs
- Witnesses
  - names
  - employers
  - statements
- Estimated cost of repair
- Recommended action
- Insurance company, policy number
- Claim? Yes/no.

Performance bonds and bank guarantees

Many contracts contain the provision of last resort – the bond. The performance bond or bank guarantee is only as good as its terms, for although the existence of a bond or guarantee may be set out in the contract, they are contracts in their own right and the use of them must be in accordance with their terms. But where they exist, they can be used by the client to pay for remedial work which the contractor is unable or unwilling to carry out. The specifier's role in calling in a bond is restricted to the technical analysis of what went wrong and how it can be rectified. Such an analysis should provide the management with the basis of calling in the bond. If suitably drafted, the bank or insurance company will not expect to see any more than the demand with reasons, and will not enquire into those reasons. However, if the bond is subsequently challenged in the courts, the reasons why it was called will have to stand up to scrutiny under the terms of the contract and the bond. The calling in of a bond is considered a fairly drastic step because it will directly affect the contractor's standing with the bank or bonding company.

Parent company guarantees

Parent company guarantees are often used by clients as an alternative source of security of performance from the contractor. This is particularly so if the contracting company is a relatively small part of a large group and if the annual turnover of that contracting company is not adequate for the size of the contract. Generally, clients are wary of awarding a contract of say more than 20% of a company's annual turnover because of the fear that the company's resources will be stretched too far. In practice, a parent company guarantee can be treated like a bond, in that the specifier will need to provide the technical analysis under the contract to decide whether or not the guarantee should be effected.

Submitting complete claims

The specifier needs to obtain from records all the evidence required to analyse or formulate a claim in one go because, in the normal course of events, it is not possible to resubmit a claim on the grounds that there is new evidence to support it once it has been rejected by the other party or, for that matter, to reject it once it has been accepted. Of course, this is not to say that the aggrieved party cannot continue to pursue the claim, but it usually means that the party will have to pursue it at a higher level, possibly through adjudication, arbitration or litigation. In either of the last two it is the common law position that all the evidence to support a case is brought forward at once in order that the matter can be decided forever. To try to appeal to the courts on the grounds of new evidence is very difficult indeed, and it will have to be proved that the evidence was unobtainable in the first instance.

Managing the claim: assembling evidence

How claims are managed is a huge subject but the specifier can play a significant part in promoting the logical and rational order of events. As important as the very existence of records is a proper investigation of them at the time the claim is being pursued or resisted. Going to the records, the specifier is interested both in establishing what was supposed to happen (in the contract documents) and in what was actually done, the latter being recorded, hopefully, in all sorts of reports, diaries, site notes, test records, and so on.

First, the specifier should classify what type of records contain the evidence for which he or she is looking and then investigate those records to find out what happened. In the example that follows the specifier found that he was being bombarded with anecdotal evidence as to the completeness of records (the problems concerned lack of data from ocean current meters) and that most of the 'evidence' was irrelevant. The contractor claimed that he was not liable as fishing vessels had removed the meters. In the end, the specifier issued the following guidelines:

Further to the discussions held recently, a few comments on future actions:

(1) No payment is due to the contractor until a final, approved report is submitted to us.

(2) When the final report has been submitted for approval, Engineering Dept should produce a comparison of what was contracted for (Attachment I to the Contract) and what was received (i.e. the volume, spread and accuracy of the current readings).

(3) Similarly to (2) above, comments are required on the Execution Proposal (Attachment IV to the Contract).

(4) Legal Dept should be asked whether (a) payment of the lump sum price depends on the receipt of 100% data *and* the final report or some other percentage *and* the final report, and (b) whether the contractor can demand full payment despite the data shortfall because of the loss/removal of meters.

(5) When the contractor has submitted his final invoice a reconciliation should be made of the above points.

(6) On the basis of the reconciliation, we should pay the final invoice less the allowable deductions.

Though apparently rather elementary, the above memorandum served to focus the engineers' attention on the real problem – the lack of data – and away from irrelevancies such as payment on account, removal of meters by fishermen, and so on. From the above can be set out a logical form that will apply in most cases:

1   What is the present position and the actual problem?
2   What remains to complete the work?
3   What was supposed to have been done?
4   What was actually done?
5   Who can advise on individual parts of the problem?
6   How to carry out the reconciliation of the various aspects?
7   How to make the commercial settlement?

Specifiers will find that most of the irrelevancies will be filtered out if they adopt the above approach (see also Importance of records, p. 184).

Money calculations

There are four possible outcomes to a claim: change in programme, modification to the constructed item or specification, monetary compensation, and combinations of all three. However, almost all claims contain an element of money, in which case the specifier's analysis should be the base on which the money calculations are carried out. With the possible exception of liquidated damages, claims do not originate by one party claiming a sum of money outright. One party comes to the conclusion that he or she has suffered some contractual wrong, and then calculates the value of that wrong – the 'value' being in recompense for the wrong suffered.

The claim must be defined. The specifier should always be wary of requests for analysis that begin 'due to X's default we have lost £Y thousands...' or 'this item will cost £Y to replace...', because it implies that the party making the request has already done the necessary analysis (unlikely), or that the party has some preconceived idea of how much has been lost as a result of a variety of problems under the contract and wants this claim to provide the compensation (very likely). In these cases the money calculations quickly develop a life of their own and become divorced from the contractual reality on which they must rely in order to succeed. It is remarkable, even now, how many claims are put forward on the basis that actual costs have exceeded tendered costs, regardless of the contract mechanisms for valuation.

The specifier, in most cases, will have to pass the calculation of the monetary effects of the claim to a specialist, probably a quantity surveyor. But in so doing, it is important to make sure that the claim remains based on the technical analysis, for the very calculation of the monetary effect can spawn further 'sub-claims'. In the claim process it is necessary to settle the liabilities first and the money second. To admit or refuse a claim on its monetary value is not part of either the specifier's or the financial specialist's role. To prevent the development of 'sub-claims', specifiers should ask to see the money calculations and to reconcile them in detail with their own technical analysis. If they do not match exactly, either by omission or addition, then they must be challenged. The specifier should remember that the financial specialists have their own priorities – which may mean that any particular claim has to contribute the maximum possible to an overall target of redress of loss.

Another part of the control process is to avoid extrapolating technical evidence to produce a much larger technical problem than was at first envisaged. While it is true that incorrect foundations will throw into question the integrity of a building, it does not follow that the whole building has to be demolished and started again. Although that is a simple example, it is meant to illustrate the fact that extrapolation from a simple beginning can (literally) bring down a whole project if it is not controlled – in this case modified and strengthened foundations may be possible. But it does occur in a small number of cases, and, oddly enough, it seems that clients fall prey to extrapolation fever, whereas contractors (probably naturally) want to contain the damage. If the process is not controlled, there is the further danger that the whole debate will sink into opinion engineering, as can be readily seen in litigation cases.

Importance of records

Despite pleas to the contrary from every quarter, the importance of records at the site is constantly overlooked. Specifiers may not be in a position to influence directly the record management on a project but they should, through requests to see and confirm actual events, ask to see all records relating to those events. This means that site diaries, photographs, test records, formal reports, level books, inspection certificates and all other papers that are (or should be) produced will need to be seen as the work proceeds. Specifiers should never allow themselves to rely on other people's recollection, but should ask for the piece of paper that confirms it. In the pursuit of any claim, accuracy is essential, for failure either to investigate the claim thoroughly at the initial stages, or to formulate it correctly as it is being put together, will inevitably lead to expense and frustration later.

In carrying out the inspection of the evidence, the specifier will also need to separate fact from fiction and unsupported contention. If, for example, it is found in a diary statement that: 'The concrete failed at 11 N mm^{-2} because too much water was added during mixing; failure should have been at 22 N mm^{-2}', then the specifier should note both the fact ('failure at 11 N mm^{-2}') and the contention ('because too much water was added during mixing'). In this case it would be necessary to find the records of the water content of the mix to substantiate the contention. Otherwise, the contention may go unchallenged for a long time, leading perhaps to all sorts of erroneous conclusions.

Programme

Performance of a contract with respect to the quality of the physical results cannot be separated from the programme of activities – time has to be a factor in quality, if only because the lack of it implies a greater planning discipline on those who carry out the work. The result is that almost any claim will have an element of planning in it. But there are also claims that arise out of the programme itself, the most basic being the failure to supply a programme at all. Most contracts insist on a detailed programme of the work. In fact, programmes are often requested with the tender. Although the programmes are usually provided by the contractor, the acceptance of them by the client puts responsibility on to the client, usually for approvals and payment. The specifier can therefore expect to be involved in planning claims stemming from the following causes:

1 failure by the client to insist on the production of the programme itself;
2 lack of reasonable substantiation that the quality of the work can be maintained or that resources are available;
3 inadequate resources on the part of the client to meet the monitoring, approval and payment requirements called for;
4 the inter-relations between the contractor and others involved in the work.

In addition, and most important if the work has been the subject of variations for any reason, is the need for the specifier to be constantly aware of how the programme is affected by such variations. Few contractors will accept that extra work can be 'squeezed in', but even if they do, the result will be shown in the programme, and the specifier may be faced with a claim which is difficult to analyse because of past acceptance of changes without their effect on the programme being adequately assessed at the time.

From the point of view of the analysis, specifiers accept, as a fact of the changing nature of the bargain between the parties to a contract, that any change in the work to be carried out allows a change in the time in which it should be carried out, unless the contract specifically states otherwise. Therefore, the specifier has to analyse every change in the programme and its effects, because although in some areas they may be minimal, some may be accepted by both sides as being neutral by way of setting off one effect against another, and the rest give rise to programme changes. The specifier should note that this applies whatever the form of payment – fixed price, lump sum or fully reimbursable. The only effect of the form of payment is on the form of monetary calculation. More rarely, the conditions of contract might force greater liabilities on one party or another with regard to changes within each party's control. Changes outside the control of the parties may be influenced by the form of payment and accompanying conditions.

In the end, the specifier must ask the following questions:

1   Did a change take place?
2   How was it authorized?
3   Did it affect the resources applied to the work?
4   Did it affect the programme duration?
5   Can it be reduced to a form that will allow money calculations to be carried out?

It needs to be borne in mind that if a change was not authorized in the manner required by the contract, the ensuing claim will have no valid contractual basis. However, this does not, in the first instance, affect the question of whether a change took place.

Prevention of claims: change orders

On the basis that changes do not happen but are created, then prevention is better than cure. Also if the parties accept that they can operate a procedure whereby changes do not become bones of contention, effort in wasted argument rather than constructive analysis will be minimized. A sensible procedure for arranging changes will incorporate at least the following:

1   means by which either party can notify the other of an impending change;
2   a system for collection and storage of relevant evidence;
3   means of logging changes;
4   means of calculating monetary and programme effects;
5   agreement that work will proceed notwithstanding failure to settle the monetary details of any particular change;
6   (possibly) an immediate reference to a third party for a decision which will allow work to proceed and arguments to be deferred.

If specifiers suspect that a contract to which they contribute includes none or few of the above points, they should explain the inevitability of changes and the dangers of not having a clear method of dealing with them.

Readiness to pursue claims

Because of the length of time over which many engineering contracts take place and the possibility of inter-related claims arising throughout the duration of the work, it is important to obtain adherence to the contract from the start, and for the specifier to indicate that reperformance of some item at an early stage of the contract is more satisfactory than trying to pursue a monetary claim at an early date. Naturally enough, most contracts start in both a state of euphoria and more or less chaos. But in the first part of the work the systems must all be set up to monitor and record and thereby provide the specifier with the information necessary to analyse claims. If, at the end of the day, specifiers are trying to convince an arbitrator that a claim should be dealt with in their favour, they will need to show that they were ready and capable of collecting the data to support the claim and actually acted on it at the time. Any person asked to rule on a claim is convinced not only by the evidence itself, but by the fact that data was collected in a systematic manner. In fact, the procedure for checking can be as valuable in proving a case, or at least in preventing it from failing, as the evidence itself.

Alternative dispute resolution

The specifier may well be involved if the resolution of a claim goes beyond the immediate players in a contract. There will quite likely be provision in the contract for others to resolve the claim, for example by expert determination or by arbitration. The specifier should be aware of how these processes fit in to the contractual and legal framework, because it may be necessary to provide factual information about the contract specification and its application in the contract in support of these procedures.

Alternative dispute resolution (ADR) is the collective term covering a variety of processes for solving contractual disputes. The mainstream dispute resolution mechanisms are arbitration and litigation. Litigation is rarely, if ever, used to solve disputes in engineering contracts, and for many years arbitration was effectively the final step in solving unresolved disputes. Arbitration is a statutory mechanism and arbitration awards can be enforced by the courts. However, arbitration has tended to become prohibitively expensive for many – a fact on which one party to a contract might have relied in trying to force the other party to forgo pursuit of a claim about some matter in dispute.

Over recent years several other approaches have developed. These have always been available and they have been used to varying degrees and effects, but today these ADR techniques are now more clearly structured in model forms of contract. The three ADR categories can be defined as:

- mediation;
- conciliation;
- adjudication.

All dispute resolution systems have a separate agenda that provides for the parties to settle their dispute amicably. This is the agenda in mediation; it is less prominent, to varying degrees, in the other forms of dispute resolution. Mediation is based on the principle of an independent third party who acts in a similar way to a counsellor. The mediator's role is to act as an outlet for each party to have their arguments heard and to help them explore the issues. Conciliation is very similar in practice to mediation. The distinction between the two which is most commonly drawn is that a conciliator acts in a partial way such that he or she can try to move the parties towards the contractually correct resolution. A mediator is more neutral; the object is to move the parties towards a resolution, regardless of whether it is a correct resolution in the context of the contract. Provisions for conciliation and mediation are sometimes made in contracts. The key point is that these procedures are not final, they can be followed by arbitration.

Adjudication is a process of resolution that is more reliant on a decision. It is similar to expert determination. Adjudication has long been a means of dispute resolution. Following the recommendations of the Latham Report (see p. 30), adjudication has been made a statutory right in construction contracts under the Housing, Grants, Construction and Regeneration Act (1996).

Paraphrasing

Claims have to be seen at many levels of management, and too often there is a general failure to present information at the appropriate level of detail.

If it is true that the higher the level of management aimed at, the less the recipient will have the time (or inclination) to read, the specifier should work at reducing the matter of a particular claim to one side of A4 paper (about 500 words at a maximum). This will entail, first, listing only what is being claimed (without specifying money, time, etc.) without any of the supporting evidence. That can then be followed by the main arguments put forward by both parties, concluding with the recommendations for the settlement. The paraphrasing may be a major task because the evidence will have involved letters, memos, photographs, measurements, drawings, plant daywork records and rates, personnel records, head office cost submissions, and so on. Therefore, the specifier must ensure that he or she has the time and resources to carry this out properly.

It is also a fact that proper paraphrasing concentrates the mind and can expose basic flaws in an argument that easily get lost in the masses of evidence. Although lawyers' pleadings are much denigrated for saying as little as possible, they cannot be written properly if the writer has a poor grasp of the facts. Developing the art is worthwhile, even if only to stop management nitpicking their way through minutiae.

Emotion

Of course, the specifier would not allow emotion to creep into the handling of claims, but it is a fact that personalities play a large part in the running of any construction site and the cross-flow of information can, and does, reflect the tensions that arise. Any analysis will involve sifting through often poorly concealed insults to get at the facts. In all writing avoid adopting a tone of personal censure. This process is made easier by referring to the parties by their corporate names or abbreviations, using the pronouns 'it' or 'their' (consistently!) and referring to people by their job titles.

Avoid emotion!

Conclusion

This final chapter about claims is a conclusion to the whole process of writing engineering specifications and the presence or absence of claims is a key indicator of how well the specification has been written. If claims do occur, a well-written specification will be a valuable aid for resolution of the dispute. However, the specifier should keep in mind the following points:

1 The work of the specifier requires technical competence and it is especially difficult when input from different disciplines has to be combined.
2 Specifications will never be 100% correct.
3 Accept that things will go wrong.
4 Strict adherence to the specification (at all costs) will lead to disputes.
5 It is all about people – a well-written specification will help cooperative people to deliver their engineering project.

Appendix A: Typical contents list for a construction (with minor M&E plant) contract

VOLUME 1

1.0 Introduction
1.1 The Documents
1.2 Early information to be provided by the Contractor

2.0 Standard Specifications
2.1 General
2.2 Interpretation
2.3 Standard Specifications
    Standard General Specification
    Standard Civil Specification
    Standard Mechanical Specification
    Standard Electrical Specification
    Standard ICA Specification

3.0 Conditions of Contract

4.0 Bonds and Guarantees

VOLUME 2

5.0 Particular Specification
5.1 Particular General Specification
5.2 Particular Process Specification
5.3 Particular Civil Specification
5.4 Particular Mechanical Specification
5.5 Particular Electrical Specification
5.6 Particular ICA Specification
5.7 List of Specification Drawings

VOLUME 3

A COMMERCIAL SUBMISSION

6.0 Pricing
6.1 Preamble

Appendix B: Typical contents list for a (process) design, construct and commission contract

CONTENTS

VOLUME 1

B   TECHNICAL SUBMISSION

15.0   Contractor's Data Sheets
15.1   General
15.2   Process
15.3   Civil
15.4   Mechanical
15.5   Electrical
15.6   ICA
15.7   Particulars of Plant and Materials to be obtained under purchasing
         initiatives – Frameworks and Best Practice
15.8   Subcontractors
15.9   Sourcing information
15.10 Erection requirement

Appendix C: Typical introductory clauses for interpretation of the specification

2.0   STANDARD SPECIFICATIONS

2.1   General

2.1.1   British Standards and other documents referred to in the Contract shall be deemed to be those current 42 days prior to the date for return of Tenders.

2.1.2   Any reference in the Contract to a Standard published by the British Standards Institution, or to the specification of another body, shall be construed equally as reference to an equivalent one.

2.1.3   In accordance with European Law, European Standards (as issued) shall have precedence over equivalent British Standards.

2.1.4   References in the Specification to the Purchaser shall be deemed to include the Employer and vice versa where either term is defined in the Conditions of Contract. (See Section 3, Volume 1 for specific definitions.)

2.1.5   References in the Specification to the Engineer shall be deemed to include the Project Manager and vice versa where either term is defined in the Conditions of Contract. (See Section 3, Volume 1 for specific definitions.)

2.2   Interpretation

2.2.1   In the event of any conflict between the provisions of these Standard Specifications (in Volume 1) and the Particular Specifications (in Volume 2), the latter shall prevail.

2.2.2   Except where the context otherwise requires, terms which are defined in the Conditions of Contract shall have the same meaning in the Specification and Schedules.

2.3   Standard Specifications

2.3.1   The following Standard Specifications are included overleaf as appropriate to this contract:

2.3.2   Reference shall also be made to the Particular Specifications and Purchaser's Data sheets in Volume 2 and to The Contract Schedules in Volumes 2 and 3.

(The reference to European Law in 2.1.3 is appropriate to a public utility company, but not generally.)

Notes

1 Specifications in context

1 Upex, R., *Davies on Contract*, 8th edn, London, Sweet & Maxwell, 1999.
2 Merna, A. and Smith, N.J., 'Privately financed infrastructure in the 21st Century', published in *Civil Engineering*, Proceedings of the ICE, London, Thomas Telford, November 1999.

2 Recent developments and standard forms

1 Barnes, M., 'Civil engineering management in the new millennium', published in *Civil Engineering*, Proceedings of the ICE, London, Thomas Telford, May 2000.
2 CE mark: the Conformité Europeène mark indicates that a product has met all EU legislation and can be sold in the EU.
3 Statutory Instrument 1996 No. 2911, The Utilities Contract Regulations 1996.

3 Presenting specifications

1 Statutory Instrument 1996 No. 913, The Offshore Installations and Wells (Design and Construction, etc.) Regulations 1996.

5 Writing specifications

1 Gowers, Sir E., *The Complete Plain Words* (Revised by Greenboum, S. and Whitcut, J.), London, HMSO, 1986.

Suggested further reading

The following list of books is suggesteed for background reading to many of the topics covered in this book. It is unusual to include a dictionary in a reading list, but it is one dictionary that is fascinating to read.

Bone, S., *Osborn's Concise Law Dictionary*, 9th Edition, London, Sweet & Maxwell, 1993.

Chappell, D., *Understanding JCT Standard Building Contracts*, 6th Edition, London, E. & F.N. Spon, 2000.

Civil Engineering Procedure, 5th Edition, London, Thomas Telford, 1996.

Cornish, W.R., *Intellectual Property*, 4th Edition, London, Sweet & Maxwell, 1999.

Cox, A. and Townsend, M., *Strategic Procurement in Construction*, London, Thomas Telford, 1998.

The Egan Report, *Rethinking Construction*, London, DETR, 1998.

Eggleston, B., *The New Engineering Contract – a Commentary*, Oxford, Blackwell Science, 1996.

The Latham Report, *Constructing the Team*, London, HMSO, 1994.

Timpson, J., Totterdill, B. and Dyer, R., *Adjudication for Architects and Engineers*, London, Thomas Telford, 1999.

Turk, C. and Kirkman, J., *Effective Writing*, London, E. & F.N. Spon, 1988.

Index

Printed in the United States
by Baker & Taylor Publisher Services